Geofuels

Energy and the Earth

ALAN R. CARROLL
University of Wisconsin

CAMBRIDGE
UNIVERSITY PRESS

CAMBRIDGE
UNIVERSITY PRESS

32 Avenue of the Americas, New York, NY 10013-2473, USA

Cambridge University Press is part of the University of Cambridge.

It furthers the University's mission by disseminating knowledge in the pursuit of education, learning, and research at the highest international levels of excellence.

www.cambridge.org
Information on this title: www.cambridge.org/9781107401204

© Alan R. Carroll 2015

First published 2015

Printed in the United States of America

A catalog record for this publication is available from the British Library.

Library of Congress Cataloging in Publication Data
Carroll, Alan R.
Geofuels : energy and the earth / Alan R. Carroll, University of Wisconsin.
   pages cm
Includes bibliographical references and index.
ISBN 978-1-107-00859-5 (hardback) – ISBN 978-1-107-40120-4 (pbk.)
1. Power resources.   2. Fossil fuels.   3. Fossil fuels – Environmental aspects.
4. Renewable energy sources.   I. Title.
TJ163.2.C385   2015
333.79–dc23   2014038370

ISBN 978-1-107-00859-5 Hardback
ISBN 978-1-107-40120-4 Paperback

Contents

Acknowledgments

The idea for this book arose from a class called Energy Resources that I began to teach at the University of Wisconsin–Madison in 2006. I quickly realized that no one instructor could ever hope to master the complexities of this important subject alone, and I therefore enlisted a number of guest lecturers, who generously agreed to help educate both my students and me. These include University of Wisconsin–Madison Profs. Tracey Holloway (climate change); Sanford A. Klein (solar energy technology); Douglas J. Reinemann and Christopher J. Kucharik (biofuels); Gerald L. Kulcinski, Michael L. Corradini, and Paul P. H. Wilson (nuclear energy and nuclear waste); and Philip E. Brown (uranium resources). These also include David Blecker of Seventh Generation Energy Systems (wind); Peter Carragher and Dr. Darrell Stanley of BP (petroleum exploration); Prof. Timothy R. Carr of the University of West Virginia (shale gas); Dr. Kenneth R. Bradbury of the Wisconsin Geological and Natural History Survey (water and energy); Dr. S. Julio Friedmann of Lawrence Livermore National Laboratory (carbon capture and storage); Dr. Philip H. Stark of IHS (petroleum resources); Dr. Paul J. Meier of the Wisconsin Energy Institute (energy planning); and Peter Taglia of Taglia Consulting (energy policy). All of these individuals helped to shape my thinking, but I am solely responsible for any errors.

I am grateful to many others for data, discussions, or other assistance, including Navin Ramankutty of McGill University (historical crop density models); Wen-yuan Huang of the U.S. Department of Agriculture (historical fertilizer data); Prof. Tim K. Lowenstein of Binghamton University (evaporite deposits); Dr. Harrison H. Schmitt (lunar ^{3}He); Prof. Shanan E. Peters, Dr. Noel A. Heim, and John Czaplewski (North American macrostratigraphy); Prof. Larry W. Lake of the University of Texas, Austin (petroleum engineering); Prof. Harold

J. Tobin and Susanna I. Webb of the University of Wisconsin–Madison (clathrate imaging); and Cara Lee Mahany Braithwait of the Wisconsin Public Utilities Institute. Dr. Meredith Rhodes Carson served as a sounding board as the book took shape and provided useful feedback. The book also benefited from many discussions that took place under the auspices of the Wisconsin Energy Institute, and the Energy Analysis and Policy program of the Nelson Institute, University of Wisconsin–Madison.

I particularly appreciate the efforts of those who provided reviews of individual chapters, including Dr. Kevin M. Bohacs (ExxonMobil Upstream Research Company); Prof. Michael E. Smith (Northern Arizona University); Eric M. Williams, Prof. Sanford A. Klein, Prof. Shanan E. Peters (University of Wisconsin–Madison), and Richard W. Crone. Dr. Matt Lloyd, Joshua Penney, and Holly Turner at Cambridge University Press were extremely helpful and patient in shepherding this project to completion. I am also grateful for the tireless production efforts of Nishanthini Vetrivel and the staff at Newgen Knowledge Works.

Prof. Stephan A. Graham of Stanford University helped to inspire this book by introducing me to integrative, "big picture" geologic thinking. Sara Godwin's inquisitive probing encouraged me to get started; I wish she could see the final product. Finally and most importantly I thank my wife, Prof. Wendy C. Crone, both for her consistent support and encouragement, and for her countless expert suggestions on how to present complex scientific topics to a general audience.

1

Introduction

Save the Planet! This slogan has been often repeated, and with good reason. The world we live on seems to be under heavy assault on numerous fronts, ranging from biological extinctions to potential shortages of key natural resources. Energy use and its consequences currently rank among the most pressing of these concerns, and not surprisingly are among the most hotly debated and divisive issues of the day. Our use of fossil fuels lies at the heart of the debate, largely because fossil fuels supply the vast majority of our current energy use (~86 percent in the United States). It has also become increasingly clear that emissions from the burning of fossil fuels are altering the composition of the atmosphere, carrying the potential for catastrophic climate change.

So how did we get to this point? We live in a world made possible by the use of fossil fuels, but it was not always so. Prior to the 19th century virtually all energy was "green" energy, mostly derived from standing biomass in the form of agricultural crops and native vegetation. This biomass could be burned directly for warmth or fed to animals and humans to produce mechanical power (the horse still represents the reference point for some power measurements today!). Most people worked in the fields, went to sleep when it got dark, and rarely traveled far from their homes. This lifestyle may have been dull, but the boredom generally did not last very long. Average life expectancy at birth was far lower than today, estimated at forty years or less. Famine and pestilence were ever-present threats, and few means were available to combat them.

Dramatic change began with the Industrial Revolution, during the late eighteenth to early nineteenth centuries. The initial rise of machines was powered largely by coal, which fueled steam engines. The origins of the Industrial Revolution are undoubtedly complex, but it clearly occurred most rapidly in countries possessing rich supplies of coal (most notably Britain, Germany, and the United States). Coal conveys significant advantages over earlier biomass fuels: It has higher energy density, its availability is not limited by arable land surface area or growing season, and it can be mined relatively cheaply. The technology originally built around coal was also responsible for the large-scale commercialization of crude oil; for example, a steam engine was used to drill the first modern oil well in the United States (operated by Edwin Drake in Titusville, Pennsylvania, 1859). Although oil and natural gas have since taken over much of the position once occupied by coal, coal-fired steam turbines continue to supply a major share of our electricity.

Today we tend to take for granted that we will always be warm and well fed. We enjoy the luxury of free time, because most farm-work and other heavy labor is done using machines. We barely think about the ease of traveling tens or hundreds of miles by automobile and cannot easily imagine living without this ability. Average people can safely and cheaply travel to another continent for a week's vaca-tion. The same journey would have cost their not-so-distant ancestors many months in travel time, a large part (if not all) of their wealth, and possibly even their lives. Average life expectancy in the United States nearly doubled during the 20th century, and world population nearly quadrupled. These increases were unprecedented in human history, and coincided with a tenfold increase in gross domestic product per capita in industrialized countries. During the same period the use of fossil fuels increased by a factor of nearly 8 (Figure 1.1). To a large extent these recent transformations were all powered by the unprece-dented bounty of fossil fuels.

It is highly doubtful that most people would want to return to the living conditions of the 18th century, even it were possible for the current population to do so. We are therefore confronted with a very challenging problem: How can we sustain the historically unprec-edented level of energy consumption of the past century and thereby continue to enjoy its benefits? Equally important, how can we extend these same benefits to the billions of people who presently consume energy at a far lower rate than those in the United States and other highly developed countries? In the past few decades it has become apparent that the "energy problem" also encompasses another dis-tinctly different but equally important question: how to avoid or reme-diate the unwanted consequences of large-scale energy use. Climate

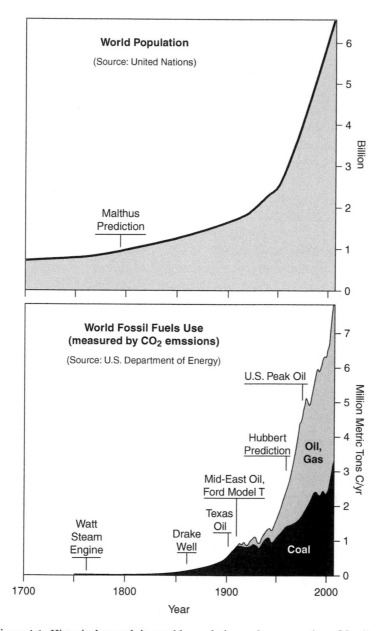

Figure 1.1. Historical growth in world population and consumption of fossil fuels, represented as carbon emissions (data sources: United Nations; Oak Ridge National Laboratories Carbon Dioxide Information Center).

changes related to the burning of fossil fuels currently loom as the most serious example of the latter. However, it is only prudent to assume that any form of energy production will incur its own unique environmental costs, especially when scaled to the present magnitude of fossil fuel use.

Just how much oil and other fossil fuels remain? It seems obvious to ask this question since we've been riding a wave of oil up to this point. Will that wave continue, or is it about to crash on the shore? Everyone knows that fossil fuels must be finite, because the planet itself is finite. A rational person might surmise that a fixed resource that is being consumed at a known rate will not last forever, and that its time of depletion should be more or less predictable. The well-publicized concept of "peak oil" is built around this reasoning, and on the assumption that the decline of oil will be a historical mirror image of its rise. Shell Oil geophysicist M. King Hubbert, who first proposed the peak oil concept, famously predicted that U.S. oil production would peak in either 1965 or 1970. In fact, it peaked in the early 1970s, emboldening others to make similar predictions of an imminent decline in world oil production (according to one such prediction world oil production should have peaked in 2005). Adding weight to the expectation of global peak oil is the decline in the discovery of giant oil fields that began in the 1970s to 1980s. The majority of oil is produced from the largest fields, so a decline in discoveries presumably signals an eventual decline in global production.

Doomsaying is perhaps among the oldest of professions, but it has not been among the most successful. The idea of peak oil has intellectual roots that go back at least as far as 1798, when Thomas Robert Malthus warned that world population growth would soon outpace growth in food production, leading in turn to widespread famine, pestilence, and poverty. This feared catastrophe failed to materialize, largely because Malthus did not anticipate the impact of the technological revolution that was already under way. Accurate prediction of oil futures has turned out to be similarly problematic, with dire warnings of shortage extending back nearly to the beginning of large-scale oil production itself. For example, global oil shortages were widely believed to be imminent during World War I, and in 1921 George Otis Smith, then director of the U.S. Geological Survey, warned, "The estimated reserves are enough to satisfy the present requirements of the United States for only 20 years." Far from running out of oil, present U.S. production rates are approximately 2.5 times higher now than they were in 1921, and in fact are about 2 times higher than predicted by Hubbert in 1956.

Can peak oil and global warming both be problems at the same time? After all, with nothing left to burn there would be no new CO_2 emissions to the atmosphere. Because fossil fuel availability and global warming represent two sides of the same coin, running out of oil might actually be a good thing for the environment. Unfortunately for the atmosphere there is not much sign of this happening yet. More ominously, no one is even talking about peak *coal,* and it appears that the known reserves are sufficient to sustain our current use rates for centuries. Recent high oil and natural gas prices have also revived interest in fossil fuel resources that are more difficult to recover but potentially very large in magnitude, such as oil sand, oil shale, and shale gas. New technologies for extracting these "unconventional" fossil fuels are evolving rapidly and dramatically altering our perception of fossil fuel reserves.

Without fossil fuels, is it even possible for us to continue living in the style to which we have become accustomed (and for the rest of the world to catch up)? This simple question unfortunately does not have a simple answer. Nuclear power currently ranks #2 behind fossil fuels, and it may be ready (like the famous car rental company) to "try harder." Like oil, nuclear power depends directly on a finite natural resource, but at first glance uranium reserves appear to be large enough to last for millennia. M. King Hubbert himself was an early supporter of the expansion of nuclear power. His 1956 paper, entitled "Nuclear Energy and the Fossil Fuels," was presented as an argument for nuclear power as a long-term replacement for fossil fuels. However, the rosy assessment of nuclear's potential presented by Hubbert (and others since) depends on counting all available uranium. Uranium in fact takes several different forms, the most common of which have atomic masses of 235 and 238. Only ^{235}U is currently used to generate power in the United States, because the reactor designs required to use ^{238}U are more costly and create enriched plutonium (which can be used in nuclear bombs) as a part of their normal fuel cycle. ^{235}U represents only 0.7 percent of the natural uranium supply however, leading to the startling implication that useful uranium reserves could become depleted within a matter of decades!

Moving further down the list (and closer to the 18th century), "renewables" are the fastest growing energy sector today. Renewable energy for the most part derives from sunlight, which can be harnessed either directly by solar collectors or indirectly by plants, moving water, or moving air. No one expects the Sun to stop shining, and so it is reasonable to believe that in addition to being renewable these energy sources will also be sustainable over long periods. In fact, these are the only energy sources for which sustainability can be historically proven; all were familiar (in primitive form) to the ancient Greeks. Renewables have never before been called upon to supply the majority of energy needs of a modern industrial society, however. Their present rapid growth is possible in part *because* their relative contribution is so small. Will this childhood growth spurt continue into adolescence and adulthood? Can the substantial technological barriers to large-scale production be overcome? What will be the environmental costs of renewable energy produced at a magnitude that replaces fossil fuels? These questions are only beginning to be explored.

So far there appear to be no clear winners for replacing fossil fuels, or at least no single winner. However, there are many smaller players that could help either to reduce the use of fossil fuels or to reduce their negative impact. We are therefore led to a diversified energy future, in which stabilization (or reduction) of CO_2 emissions is an overarching objective. This diverse future can be represented graphically by "stabilization wedges," as originally proposed by Socolow and Pacala in 2004 (Figure 1.2). Each wedge represents a different, partial solution to the larger problem of CO_2 emissions. The wedges can represent new alternative energy sources, reductions in net release of CO_2 from fossil fuels, or more efficient energy use. The relative sizes of the wedges vary, depending on who does the projecting, but most projections start with the assumption that future world energy demand will continue to grow geometrically at rates similar to those of the recent past.

But are all wedges created equal? Presumably not; each wedge has its own particular benefits and costs, which can be both economic and environmental. Some proposed solutions could even turn out to be more harmful in the long run than the problem they are intended to fix. The sheer complexity of scientific, economic, and political issues created by a diversified energy portfolio can be dizzying. The ongoing debate over corn ethanol, which presently consumes roughly 40 percent of U.S. corn production, serves as an excellent case in point. Some detractors claim that it is not an energy source at all, because the energy consumed in its production exceeds the energy content of the resultant fuel. Proponents have argued otherwise, concluding that for every 1 unit of energy invested, perhaps 1.5 units of energy are returned. This is a good thing, right? Maybe. Good, in that it adds 50 percent leverage to

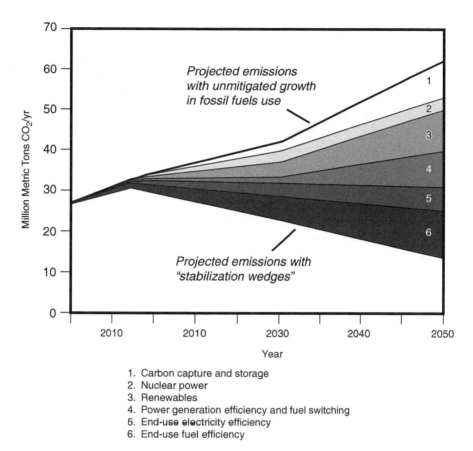

Figure 1.2. CO$_2$ stabilization strategies (modified from International Energy Agency 2008).

energy obtained from fossil fuels, possibly resulting in lesser CO$_2$ emissions. Also good if corn ethanol helps to build a technological bridge to superior cellulose-derived fuels, which are expected to have better energy returns. Bad however, in that corn ethanol and other biofuels consume huge amounts of water, require extensive use of fertilizers, and may promote soil erosion. A 1.5:1 energy return ratio means that about three acres must be put into production to recover the amount of solar energy captured by one acre of corn, effectively multiplying any environmental consequences threefold.

Confused yet? The slopes only get slipperier when economic and political considerations are applied. Are corn and ethanol subsidies vital to support American farmers, or are they a giveaway to big agribusiness? Are biofuels an important step toward reducing our dependence on foreign oil, or do they just drive up food prices? All of these views (and more) have their proponents, none of whom is shy about speaking up. The same can be said for most other energy systems, including fossil fuels. In some cases the advocates may be speaking from a genuine belief that their solution is in the best interest of the largest number of people. In other cases the motivation probably derives from more mundane self-interest. It can be truly difficult to know whom (if anyone) to believe, and the physical, economic, and political realities of energy have become entangled into a kind of Gordian knot. According to legend, Alexander the Great sliced apart the original Gordian knot with a stroke of his sword, and from there went on to rule over most of the known civilized world. Regrettably, solving the energy problem will probably not be so easy as Alexander's conquests.

Before we can even hope to conquer the energy problem we need to understand it better. The Earth itself can help in this regard. Everyone can agree that Earth resources such as fossil fuels are finite, and that the long-term history of the Earth over millions of years has determined their availability. There is plenty of room for debate over the precise natural abundance of such resources, how long they might last, or whether we should really be using them at all. However, we can at least point to some concrete observations that are not likely to change and that are not really subject to debate. For example, fossil fuels are not renewable over time frames that humans normally think about. The discovery rate of giant oil fields is declining, but production of "unconventional" oil and natural gas is increasing. Burning of fossil fuels results in relatively rapid release of carbon that originally required millions of years to remove from the atmosphere. Such observations help to illuminate and clarify related energy issues in terms of physical constraints imposed by the natural history of the Earth.

Similar logic can be applied to virtually all energy systems, because all consume natural resources in one way or another. This may sound like a rather radical statement; doesn't renewable by definition mean *not* dependent on finite natural resources? Yes and no. The energy itself is certainly infinite for practical purposes, assuming the Sun doesn't burn itself out any time soon. However, the means of gathering and using this energy generally are certainly not infinite. For example, the soil and water needed to grow crops for biofuels are both limited natural resources. Soil quality has already been degraded by agriculture in many parts of the world (Figure 1.3), and once gone it cannot be quickly restored. Rainfall continually replenishes surface water supplies, but agriculture in areas such as the Great Plains of the United States relies heavily on irrigation by groundwater. The

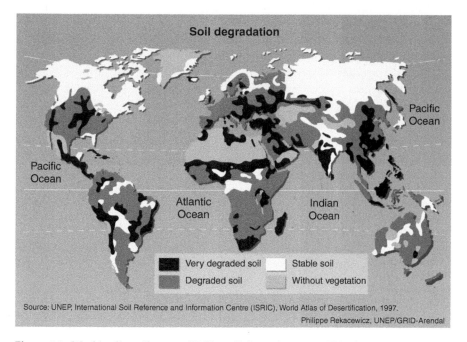

Figure 1.3. World soil quality map (Philippe Rekacewicz, 2007, UNEP/GRID).

major aquifer there is being rapidly depleted and may require thousands or tens of thousands of years to naturally recharge. Even in areas with plentiful rainfall, fresh water supplies are often considered an endangered resource because of competition between urban and agricultural uses, the environmental impacts of water usage, and contamination.

Geology also exerts first-order control on the availability of a variety of other renewable resources. For example, regional patterns of sunshine and wind intensity are governed in part by the geometric configuration of the continents, which profoundly influences the circulation of the ocean and atmosphere. The geologic history of the continents has also determined the fine-scale availability of wind resources, as a result of the dependence of wind resource quality on topography. Even tidal energy is dependent on geologic history, because effective use of tidal energy requires favorable coastline geometries to amplify tides. Moving deeper, the connection of geology to geothermal energy should be obvious in places with active hot springs. Less obvious is the fact that geothermal energy systems often rely on natural groundwater circulation that can take hundreds or thousands of years to renew.

Earth resources also govern our ability to dispose of the by-products of energy production. Practically speaking, only three places are available to dispose of things we don't want: the atmosphere, the ocean, and the solid Earth. If we don't want to put the CO_2 released from fossil fuels into the first, we must then consider one of the other two. As it happens, the ocean is already absorbing about one-third of anthropogenic carbon emissions. It has been suggested that this fraction could be increased by either stimulation of uptake of CO_2 by organisms in surface waters or by direct injection of CO_2 into deep waters. Neither of these approaches has yet been fully explored, and the potential for unanticipated consequences is largely unknown.

In contrast, industrial-scale injection of CO_2 into the solid Earth has been routinely practiced for decades. Ironically, the purpose has not been to reduce atmospheric greenhouse gases, but instead to stimulate greater rates of oil production. No apparent ill effects have resulted so far, and the CO_2 appears to stay buried. This experience has lent encouragement to the idea of larger-scale injection of captured CO_2 from power plants or other point sources. This is essentially petroleum geology in reverse, returning exhumed carbon to its original resting place. Whether geologic carbon storage can be eventually deployed on the massive scale required to have a noticeable effect on the atmosphere remains to be seen. Several pilot projects and a considerable amount of research are currently aimed at addressing questions of reservoir capacity, protection of groundwater resources, leakage back to the atmosphere, and cost.

Underground storage may also be the preferred option for dealing with the by-products of nuclear power generation. The basic requirement for geologic disposal seems relatively simple: long-term isolation of nuclear waste from the surface environment and from potential water supplies. The implementation of underground waste repositories has been anything but simple, however, because of a unique combination of political and technological challenges. For example, the Yucca Mountain site in southern Nevada was designated as a future national waste repository in 1987 but met strong local opposition from the start. The project was finally cancelled in 2009 after an estimated $30 billion in research and development expenditures over a period of more than twenty years.

One of the more challenging technical requirements of the Yucca Mountain site was that it maintain a specified level of integrity over time periods of up to 1 *million* years. To put this time interval in perspective, the pyramids of Egypt are only about five thousand years old. Twenty thousand years ago the geography of North America looked radically different than it does today. Glacial ice sheets up to

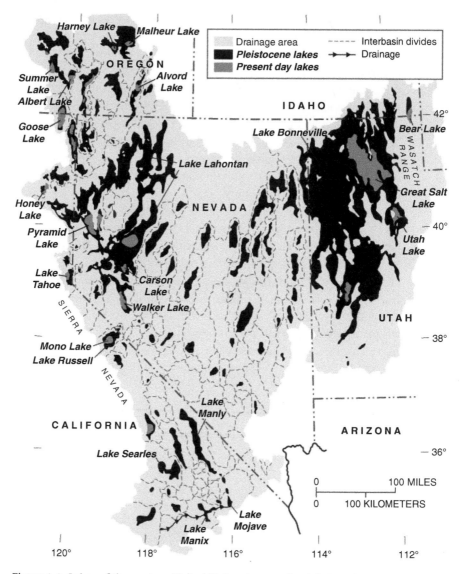

Figure 1.4. Lakes of the western United States at approximately twenty thousand years ago (U.S. Geological Survey).

two miles thick covered all of Canada and parts of the northern United States, including what are now the cities of New York, Boston, and Chicago. The present desert landscape of Nevada and Utah was then a relative paradise, featuring lush vegetation and dozens of large fresh-water lakes (Figure 1.4). *Homo sapiens* (modern humans) only appeared about 200,000 years ago, and it seems optimistic to think we will still be around in another 200,000 years. In the greater scheme of things, planning 1 million years into the future lies somewhere between hubris and sheer fantasy.

The following chapters explore in more detail what Earth history can tell us about both the availability of energy and the consequences of its use. We'll start with an introduction to the Earth itself. Far from being merely a pile of old rocks, it is actually quite a lively planet. In fact if the Earth could speak, it would probably express bemused surprise at our concern for its safety. It has, after all, survived for 4.54 *billion* years so far and will most likely continue to circle the Sun for at least a few more billions of years. The Earth has survived many previous crises, the history of which is literally carved in stone. It has been repeatedly pummeled by meteorite impacts, wracked by earthquakes and volcanic explosions, and baked or frozen during countless climatic fluctuations. It has witnessed the diversity of living organisms rise, fall, and rise again in new forms, dozens of times. Through all of this the Earth itself has proven quite durable. It is of course not really the planet that needs salvation, but ourselves.

For More Information

International Energy Agency, 2008, Energy Technology Perspectives: Scenarios and Strategies to 2050, Executive Summary: Paris, France, Organization for Economic Cooperation and Development/International Energy Agency, 12 p.

Pacala, S., and Socolow, R., 2004, Stabilization Wedges: Solving the Climate Problem for the Next 50 Years with Current Technologies: *Science*, v. 305, p. 968–972.

2

The Living Earth

The result, therefore, of our present enquiry is, that we find no vestige of a beginning – no prospect of an end.

James Hutton, 1788

As planets go, the Earth is a party animal. Its continents wander endlessly, sometimes clustering together in temporary cliques, other times splitting up and disbanding. Their surfaces may plunge 10 km down or more, buried beneath thick blankets of sediment, or skyrocket 8 km upward, shaking violently the whole while. Fireworks are on perpetual display in the form of volcanoes, which emit anything from ropy lava flows to giant dust clouds that blot out the Sun. Glacial ice appears, disappears, and returns again with alarming speed and frequency. The sea rolls relentlessly in and out, creating beachfront property in St. Louis or land bridges from Alaska to Asia.

So why don't we notice? Sometimes we do, particularly during major earthquakes and volcanic eruptions. Generally though we experience an illusion of stability, because we are relatively short-lived animals inhabiting a very long-lived planet. Most of the action happens too slowly for us to perceive directly. To get a better sense of the vast discrepancy between geologic and human timescales, it helps to compare the history of the Earth to a single calendar year. Marked on January 1 of this calendar is the notation "Earth condenses from the presolar nebula." The earliest life-forms appear sometime in late March, but complex animals do not evolve until November 11. The dinosaurs appear on December 10 and disappear the day after Christmas. Modern humans (*Homo sapiens*) show up on New Year's Eve and are quite late to the party, arriving only twelve minutes before midnight. The entire 20th century is encompassed in the last tick of the second hand before midnight.

The Birth of the Earth

The Earth originally condensed about 4.55 *billion* years ago, a time so distant that even the Earth itself can't easily recall the details. No intact rocks from the Earth's first ~500 million years have been found, leaving a substantial gap in the early record of the planet's evolution. Geologist Preston Cloud proposed in 1972 to name this rockless interval the Hadean, from a Greek root meaning "unseen" (Hades was also the Greek god of the underworld). Whatever happened in the Hadean, stayed in the Hadean.

Unseen doesn't necessarily mean unknown, however, and through some clever detective work the earliest events in Earth can be reconstructed with varying degrees of confidence. We also know that another large celestial body struck the Earth not long after its formation, resulting in the separation of the Moon. The energy of this impact would have been enough to briefly vaporize a large part of Earth's original rocks, but rapid loss of heat back to space allowed them to recondense. For a time the Earth consisted of a round ocean of magma (molten rock), with only a thin rock rind at its surface.

As the rock-forming elements condensed into liquid and eventually solid form, they left behind an atmosphere rich in carbon dioxide, and by 4.4 billion years ago enough liquid water appears to have accumulated to form a primordial ocean. The Earth continued to lose its original heat of formation into space, and the ocean might have actually frozen solid had it not been for the presence of a thick, greenhouse atmosphere that helped to retain heat. The Earth and other planetary bodies also continued to experience heavy bombardment by space debris left over from the early formation of the solar system up until about 3.8 billion years ago.

The first signs of biologic life appeared around the time the bombardment ended, in the form of primitive microorganisms. Microbes began to convert substantial amounts of carbon dioxide into oxygen starting about 2.5 billion years ago, resulting in the deposition of extensive deposits of iron oxides, the mineral equivalent of rust. A second rise in atmospheric oxygen occurred between about 1 billion and 541 million years ago, setting the stage for evolution of more complex organisms.

What Lies Beneath: The Inner Workings of the Earth

It may easily be deduced that the Earth has a complex internal structure simply by knowing its size and mass. The former quantity was estimated with surprising accuracy by the ancient Greek astronomer Eratosthenes, on the basis of measurement of the angles at which the Sun's rays struck the Earth at two different locations at the same time (he assumed the rays to be parallel to each other). The Earth's mass was determined by the English scientist Henry Cavendish in the late 19th century, on the basis of the strength of its gravitational attraction. If you divide the Earth's mass by its volume you obtain its average density, which works out to be about 5.5 grams per cubic centimeter. Strangely enough, this calculation conflicts with measurements of the density of actual rocks collected near the Earth's surface, which range between 2.5 and 3.0 g/cm³. The apparent mismatch is reconciled by the realization that the Earth's interior must contain materials with densities *greater* than 5.5 g/cm³, which in combination with the less dense rocks at the surface give the correct average density.

It can also be easily deduced that the Earth's density variations are arranged in regular concentric layers (Figure 2.1); otherwise the interior mass imbalance would quickly tear apart the rapidly rotating Earth. These layers nest one inside the other, not unlike the famous wooden dolls that you take apart to find another doll inside, and another inside of that. Unlike the dolls, however, the Earth's layers become progressively denser as you move toward its center. The heaviest chemical elements sank downward when the Earth was still effectively a ball of magma. The Earth's inner core is composed mostly of the solid iron and nickel and reaches densities approaching 13 g/cm³. The inner core is surrounded by an outer liquid core that is of similar metallic composition. Movements of the liquid outer core caused by the Earth's rotation are responsible for generating the magnetic field, without which we would all have been fried long ago by solar radiation. Together the inner and outer cores compose a bit more than half of the Earth's diameter and about 15 percent of its volume.

Most of the Earth's volume (about 83 percent) is contained in the next layer out, called the mantle, which is composed mostly of the lighter elements oxygen, silicon, and magnesium. The oxygen and silicon atoms combine in the ratio of 4:1 to make three-dimensional shapes called silica tetrahedra. These tetrahedra combine with each other or with magnesium and iron to form "silicate" minerals such as olivine and pyroxene. Although the mantle is technically solid, it is also very hot, and over periods of millions of years it can flow like warm road tar. In geologic terms the mantle really is the "life of the party": Enormous volumes of rock constantly migrate thousands of kilometers upward and downward in a never-ending attempt to erase

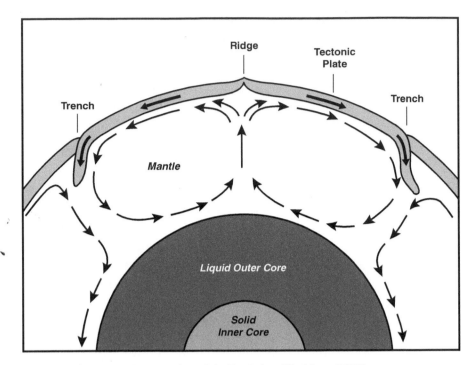

Figure 2.1. Cross section of the Earth (modified from USGS).

the temperature difference between the Earth's core and its surface. This inner turmoil sets Earth apart from the other solid planets of the solar system, which are largely dead by comparison. It is also directly or indirectly responsible for more attention-getting surface instabilities such as earthquakes and volcanoes.

Thin Skin: Earth's Moving Plates

The uppermost layer of the Earth consists of relatively thin tectonic plates, which float atop the deeper layers much like ice floating on the surface of a pond. These plates are comparatively cold and rigid, a handy feature for humans and other organisms that might otherwise be incinerated by the hot mantle rocks underfoot. The plates range in thickness from about 15 to 200 km, which represents only about 0.2 to 3 percent of the depth to the center of the Earth. The plates actually consist of two different materials laminated together: a lower part consisting of a thin slice of the mantle, and an upper part consisting of less dense rock of the Earth's crust. The crust beneath

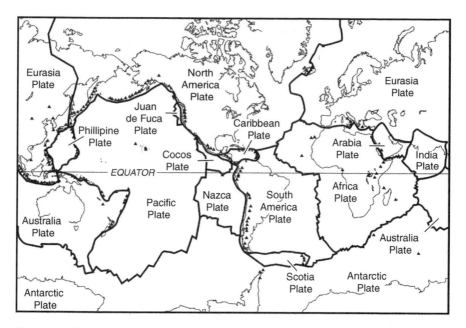

Figure 2.2. Earth's plate boundaries (modified from USGS).

the oceans averages only about 7 km thick, a distance that a reasonably fit person could jog in an hour. The continental crust is thicker, about 35 km on average, and generally contains rocks less dense than those beneath the ocean. The continents therefore can be thought of as being a bit like icebergs, floating higher than the surrounding ocean basins.

The tectonic plates are in continual motion, sliding on a relatively thin layer of mantle rocks that are partly molten, called the aesthenosphere. As they move, their edges bump and grind against each another in one of three basic ways: The plates can move toward each other head-on, they can move away from each other, or they can slide by each other sideways (Figure 2.2).

The outcome in the head-on case depends on the type of crustal rocks carried by each plate. If one of the colliding plates' edges is topped by ocean crust, it can dive submissively beneath the edge of another plate in a process called subduction. The Pacific Ocean basin is presently doing just that along most of its periphery, subducting beneath South and Central America in the east, Alaska in the north, and Asia in the west. The Pacific is therefore getting smaller through time, although not at a rate that will require any near-term revision of airline schedules. Rates of plate movement are on the order of a few centimeters per year, comparable to the rate at which your fingernails grow.

The fate of a subducted plate is to continue sliding downward into the depths of the mantle, but it doesn't go quietly. The entire circum-Pacific region is sometimes referred to as the "ring of fire" because of its proclivity for earthquakes and volcanoes. The earthquakes result from the fact that subduction is not really a smooth and continuous process, but instead chatters along by fits and starts. Volcanoes form as the descending rocks are heated up and begin to melt, sending hot magma upward into the overriding plate. The Andes Mountains in South America are dotted with volcanoes that form where this rising magma pierces the surface. As the magma cools, it forms igneous (from a Latin root meaning fiery) rocks.

At the other extreme is Iceland, an island nation that straddles two different tectonic plates that are moving directly away from one another. In fact, the entire Atlantic Ocean is expanding at the same time that the Pacific is shrinking, and has been doing so for about 180 million years. Originally the Americas, Eurasia, and Africa were all united within a single megacontinent called Pangea, which subsequently split or rifted apart to form what is now the Atlantic Ocean. The boundary between the western and eastern sides of the Atlantic is a submarine mountain range that runs roughly north-south. At this boundary the hot rocks of the mantle rise very close to the sea bottom. As they do so they experience a decrease in pressure, which allows them to partially melt. The resultant magma forms a continuous line of volcanoes called an oceanic spreading center, where new ocean crust is formed. On a global scale this process balances the loss of oceanic crust to subduction; otherwise the mantle would eventually go bald.

The third option is for two different plates to slide by each other sideways, no questions asked. One of the best examples of this type of plate boundary is the famous San Andreas Fault in California. The fault divides the North America Plate from the Pacific Plate, with the Pacific moving to the northwest relatively speaking. Los Angeles mostly lies on the Pacific Plate and San Francisco mostly lies on the North America Plate; at present rates of plate movement the two cities should become tectonically conjoined in only another 16 million years. The migration will be anything but peaceful, marred by frequent earthquakes. The San Andreas is also surrounded by numerous other faults that are related to the same plate movements, and that generate earthquakes of their own. Volcanoes are not part of this picture, however, except in overzealous Hollywood dramatizations.

The continual destruction of tectonic plates by subduction, combined with the generation of new plates by spreading, means that the Earth undergoes a continual facelift. About 70 percent of the planet is covered by water, and the *oldest* surviving ocean crust formed only about 180 million years ago. A primary reason the Earth shows so few signs of aerial bombardment by space debris is that most intense bombardment occurred long before the present oceans formed. The Moon, on the other hand, is a sort of *Portrait of Dorian Gray*, showing the full ravages of time on its deeply cratered surface. In contrast to the Earth, it has no active mantle, no plate tectonics, and no atmosphere to resist even the tiniest meteorite.

Ups and Downs: Mountain Ranges and Sedimentary Basins

Compared to the ocean basins the continents are shockingly old; the oldest intact rocks date back nearly 4 billion years. The continents owe their antiquity largely to the fact that continental crust is made of thicker, less dense stuff than the oceanic crust. As a result, the continents strongly resist subduction into the mantle. In fact, they have actually grown larger through time through the gradual accretion of island chains, marine sediment, and other low-density rocks. The oldest rocks tend to be found within the cores of continents, with younger rocks plastered around their edges.

Despite their greater thickness, continents are somewhat weak and flabby compared to younger, thinner, and stronger oceanic crust. Plate movements and other tectonic forces can therefore squeeze or stretch continental rocks at will, dramatically deforming them into new shapes and creating faults and folds in the upper part of the continental crust. Faults are simply fractures within the crust, along which the rocks on either side have moved during their deformation. Folds are similar in appearance to wrinkles in a scrunched-up carpet. Although rock layers may appear quite solid where they crop out, they can be folded if exposed to high temperatures and pressures for sufficient lengths of time.

To a first approximation, squeezing continental crust horizontally causes it to thicken and rise vertically, because the total volume of available rock remains largely unchanged. The same basic principle can be easily illustrated by squeezing the edges of a block of modeling clay. Thickening of the crust is the principal (although not the only) way in which mountain ranges form. Temperature and pressure of the crust both increase as it thickens, permanently transforming its rocks in the process. Rocks that have experienced such trauma are therefore given a new name, metamorphic, from Greek roots that mean "after" and "shape." The most dramatic example of continental thickening presently is found in the Himalaya Mountains and adjacent Tibetan Plateau. The crust in this region is about twice as thick as normal, a consequence of collision between India and the rest of the Asian continent that started around 50 million years ago.

Stretching of continental crust typically has the opposite effect of squeezing, causing it to become thinner and lower in elevation. Stretching is often concentrated in relatively narrow zones, and taken far enough this process can actually split a continent in two. It is no accident that the shapes of North and South America's coastlines mirror those of the opposing European and African coasts. At one time what is now Nova Scotia was located only a short distance from Morocco. An early stage of continental splitting or "rifting" can be seen today in the Red Sea region, where Saudi Arabia is splitting apart from Africa.

Rocks that are uplifted in mountains tend to become heavily fractured, breaking them into smaller bits that are further degraded by interaction with air, water, and organisms. The net result is that mountains are gradually reduced into gravel sand, mud, and other sediment, which is carried away by rivers. Rivers continue to flow until they enter the ocean or a lake, at which point their job is done and they can carry their load no farther. They dump large quantities of sediment into the seas immediately adjacent to the continents, ultimately forming thick, layered accumulations within what are known as sedimentary basins. For example, the entire Gulf Coast region of the southern United States is one large sedimentary basin, formed where the Mississippi and other rivers enter the Gulf of Mexico. Sedimentary basins are major first-order features of the Earth's crust (Figure 2.3). Gulf of Mexico sedimentary deposits at their thickest extend downward more than 15 km, equal to about eight times the depth of the Grand Canyon or twice the height of the world's tallest mountains!

Sedimentary basins contain virtually all of the coal, oil, and natural gas resources of the world, along with much of its freshwater and commercial deposits of uranium. The layers of sedimentary rock are also directly analogous to pages in a book, in which is written the geologic history of the Earth. Although some of the largest basins are found at the boundaries between continents and oceans,

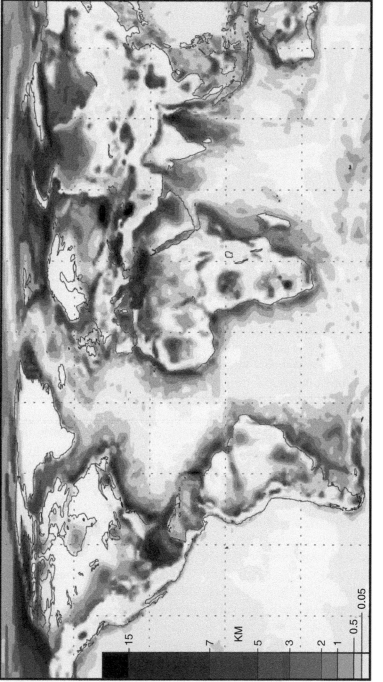

Figure 2.3. Thickness of sedimentary rocks on Earth; gray areas correspond to sedimentary basins (darker = greater thickness) (modified from Laske and Masters, 1997).

the relationship is not exclusive. Sedimentary basins can also form far inland, in response to continental stretching and squeezing and to dynamic movements of the underlying mantle.

Telling Time

To appreciate the ever-changing nature of the Earth requires the ability to comprehend events that unfold over timescales that vastly exceed the range of human experience. The Earth itself has helped us out, however, by conveniently arranging sand, mud, or other sediments into horizontal layers that become sedimentary rock. The utility of this arrangement derives from the simple principle that layers near the bottom of the stack must have been deposited first, and that those near the top were deposited last (Figure 2.4). Even if the layers are later displaced from their original horizontality by tectonic forces, their relative spatial sequence remains unchanged. This basic truth allows us to visualize the progression of geologic time in a purely *relative* sense, without the need to struggle with unimaginably long stretches of absolute time.

The different rock layers also contain distinctively different groups of fossils, and these fossil groups change progressively and irreversibly as you go upward through the layers. By comparing fossils found at different locations it is therefore possible to trace rocks of similar age across the entire planet. The fossil-bearing layers also allow the spatial distribution of different environments, for example, river plains or shallow seas, to be reconstructed for a single moment in geologic time. By flipping through a series of such maps quickly one can see how the face of the planet has changed through time. When so viewed, the hyperactive behavior of the planet becomes impossible to ignore.

Through a geologic twist of fate that is only beginning to be understood, only the most recent one-ninth of Earth history is uniformly and consistently recorded on the continents by layered sedimentary rocks. The oldest of these rocks are named Cambrian after the Roman name for Wales (Cambria), where they were first described in detail. Cambrian rocks in many areas of the world rest directly upon much older igneous or metamorphic rocks, leaving a time gap in the geologic record that may easily span a billion years. The surface marking such a gap is known as an unconformity, and the one at the base of the Cambrian has been labeled the "great unconformity" out of respect for its extreme magnitude. Rocks occurring beneath the great

Figure 2.4. Layered sedimentary rocks in Utah.

unconformity are generally more cryptic and less completely understood than those above. In the geologic timescale they are somewhat
crudely lumped together as Precambrian.

As it happens, fossils of complex animals also appear in abundance for the first time in Cambrian rocks, and continue to evolve in
overlying layers. This combination of layered sedimentary rocks and

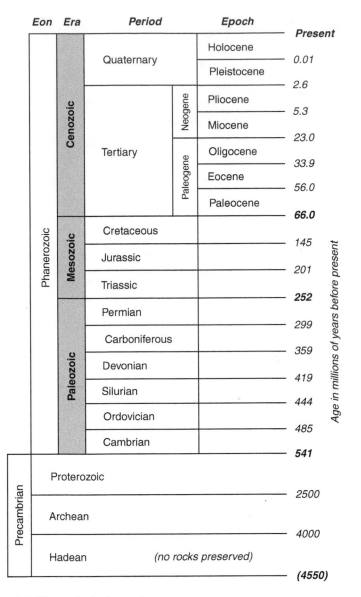

Figure 2.5. The geologic timescale.

their included fossils has led to the development of a highly sophisticated geologic timescale, which underpins all of our understanding
of the Earth's evolution (Figure 2.5). Geologists use this timescale to
describe the relative ages of different rock layers. Cambrian and younger rocks are collectively labeled Phanerozoic because they contain

visible evidence of life (from the Greek roots *phaneros*, meaning visible, and *zoion*, meaning animal). Phanerozoic time is subdivided into Paleozoic, Mesozoic, and Cenozoic eras, using prefixes that mean old, middle, and new, respectively. Each of these is further sudivided into periods, with names that commonly refer to locations where such rocks are found. For example, the Jurassic is named for the Jura Mountains in France and Switzerland.

Earth's Breath: Rocks, Water, and Climate

The Earth's dynamic mantle continuously reshapes the surface of the Earth from below, by driving the movement of tectonic plates. A different kind of reshaping occurs from above, caused by the actions of water, atmospheric gases, and sunlight. Water and atmospheric gases are not completely independent players, however; they are themselves closely tied to the ongoing movement of tectonic plates in a way that might be compared to the respiration of a living organism. The Earth effectively inhales water and atmospheric gases (in mineralized form) wherever water-saturated, sediment-laden oceanic plates are subducted into the mantle. Volatiles are also continuously exhaled from volcanoes that form above subduction zones and at oceanic spreading centers.

On average the net balance of water in the ocean appears to have remained relatively constant over the past 541 million years, maintaining an average water depth of about 4,000 m. The amount of water in the ocean generally has remained fairly well matched to the size of the basin that holds it, but ocean depth has not remained entirely stable. Instead sea level has continuously oscillated up and down, over a range of a few hundred meters. The timing of these oscillations ranges from tens of thousands to hundreds of millions of years.

The cause of some of these fluctuations is clear: the formation and subsequent melting of continental glaciers. For example, the glacial advances of the most recent ice age drew the ocean down by about 100 meters compared to today, exposing marginal areas of the continents that today are flooded (called continental shelves). Complete melting of all present glacial ice, an admittedly unlikely event, would cause sea level to rise by approximately 70 m and inundate most of the world's largest cities. Past sea level fluctuations appear to have drastically affected the evolution of organisms, particularly those that live in shallow marine environments. One recent study by University of Wisconsin-Madison professor Shanan Peters directly implicates changing sea level as a cause of repeated major extinctions over the past 541 million years.

Like the ocean, the major constituents of the Earth's atmosphere, oxygen and nitrogen, have remained tolerably stable over the past 541 million years, but are not entirely static. The compositions of air bubbles preserved in amber from the end of the Cretaceous period reveal that oxygen made up as much as 35 percent of the atmosphere, just prior to the extinction of the dinosaurs. Today it is down to 21 percent.

Carbon dioxide and other trace gases can vary much more dramatically, in part *because of* their low concentrations. Relatively small additions or subtractions of trace gases can greatly alter their relative abundance. For example, preindustrial CO_2 is estimated to have made up about 280 parts per million (ppm) of the atmosphere, or 0.028 percent, which certainly looks like a small number. Humans have already managed to increase it by nearly 40 percent to 400 ppm, with most of this change attributable to increased burning of fossil fuels over the past sixty to seventy years. Small does not necessarily mean unimportant, however, and past natural CO_2 fluctuations have been associated with dramatic changes in the Earth's climate.

CO_2 is well known as a "greenhouse" gas, which functions similarly to the glass in an actual greenhouse. Visible light, radiated by the Sun at nearly 5800°C, passes through the Earth's atmosphere relatively unimpeded and is absorbed by ground below. This sunlight warms the Earth's surface, which in turn radiates a warm glow of its own. However, because its temperature is much lower than the Sun's, it radiates energy at a much longer wavelength. Some of this infrared radiation from the Earth's surface is absorbed by CO_2 in the atmosphere, causing the air to warm.

Greenhouse gases generally are a good thing; Without them the average temperature on Earth would likely drop to a frigid −20°C and life as we know it would largely cease to exist. Too much of a good thing can be bad, however. As humans what we really want is for CO_2 to remain stable, so that we can continue to live in the manner to which we have been accustomed. However, the record preserved in sedimentary rocks shows that past atmospheric CO_2 has been anything but stable. At one extreme it has reached levels that even the most pessimistic climate scientist would never predict, resulting in sweltering conditions. At the other extreme, ice ages have gripped the Earth for periods of tens of millions of years. It has even been proposed that the Earth's entire surface was essentially

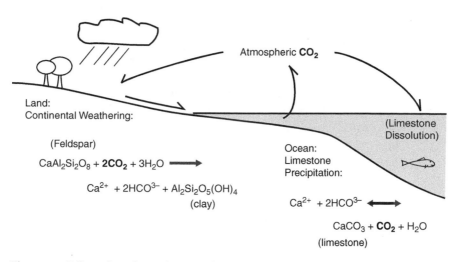

Figure 2.6. Effect of continental weathering cycle on atmospheric CO_2, based on a representative feldspar dissolution reaction.

The net result of increased weathering is to bury CO_2 in limestone deposited on ocean floor. Note that increasing atmospheric CO_2 drives the reaction in the ocean backward, dissolving limestone and storing carbon as HCO^{3-} ions (bicarbonate) in seawater.

frozen solid at least twice in its history, between approximately 700 and 600 million years ago!

As wild as these gyrations have been, the Earth itself imposes certain limits on atmospheric CO_2 concentration, which help gently nudge the system back toward a moderate middle ground. These feedbacks might be compared to a highway with gently rising embankments on either side. If a driver falls asleep and wanders off the road, the embankments tend to push the car back toward center, hopefully without hitting any other cars in the process. A driver who survives such an experience is guaranteed to be wide awake, well frightened, and extra careful in the future. In technical terms this is known as a negative feedback, meaning that it acts against change.

Rising atmospheric CO_2 triggers a negative feedback due to its tendency to combine with water to form a weak acid that is corrosive to rocks (Figure 2.6). CO_2 is consumed in the dissolution of rock-forming minerals on the continents and then carried into the ocean by rivers. Once there, it combines with calcium to form the mineral calcite ($CaCO_3$), the main constituent of limestone. Some carbon is converted back to CO_2 in the process, but for every *two* molecules of CO_2 that were originally consumed, only *one* molecule is released. The other molecule is buried as limestone. Rock dissolution and limestone precipitation both tend to accelerate when global temperatures rise, removing CO_2 from the atmosphere and thereby limiting greenhouse warming.

At the same time that mineral dissolution removes CO_2 from the atmosphere, the Earth is also pumping it back. Limestone deposited on oceanic crust is eventually carried to subduction zones, which in turn release CO_2 through volcanic eruption. Limestone buried deeply within the roots of mountain ranges or at the bottom of exceptionally thick sedimentary basins can suffer a similar fate, allowing CO_2 to seep back to the surface. These processes complete the cycle that started with mineral dissolution and help prevent the Earth from freezing over, mostly. If atmospheric CO_2 decreases in concentration, mineral dissolution and limestone burial slow down, until natural degassing of the Earth's interior eventually builds CO_2 concentrations back up again.

From the preceding discussion it may seem that plants and other living organisms are just spectators of the Earth's natural respiration of carbon, but actually they are intimately involved. Plants accelerate the rate at which rocks are broken down, by helping physically to dislodge rock particles and by contributing organic acids that speed up mineral dissolution. Plants also remove large amounts of CO_2 from the atmosphere during photosynthesis. Most of this CO_2 is released back to the atmosphere after the plants die and decay, in which case it has no net effect on greenhouse warming. However, if instead the dead plants become buried in soils or sediments, their carbon is effectively removed from contact with the atmosphere. Such burial happens primarily in sedimentary basins, as part of the ongoing accumulation of thick layers of sedimentary rock. Burial of dead plants has profoundly influenced the evolution of Earth's climate and is also responsible for essentially all of our fossil fuel resources.

A Brief History of Carbon

Extreme greenhouse climate prevailed on Earth 500 million years ago, with estimated atmospheric carbon dioxide levels in the range of fifteen to twenty times their present concentration (Figure 2.7). However, burial of organic carbon progressively increased with the evolution of vascular land plants and reached a peak about 300 million years ago. Simultaneously, continents that had previously wandered alone began to clump together, eventually forming the supercontinent Pangea.

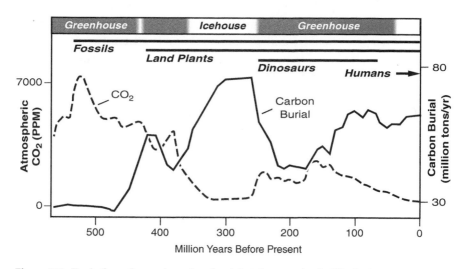

Figure 2.7. Evolution of organic carbon burial and atmospheric CO_2 during the past 543 million years (modified from Berner, 2003).

Low areas of Pangea subsided in large sedimentary basins that hosted coal swamps, burying huge amounts of organic carbon. Coal deposits in the eastern United States and in England were formed during this period, which is known as the Carboniferous. At the same time, erosion and chemical dissolution of Pangea's mountains helped to reduce atmospheric CO_2, by effectively burying carbon at sea. The combined burial of organic and inorganic carbon helped to transform the earlier greenhouse into an icehouse climate, leading to the episodic advance and retreat of glaciers across southern Pangea for about 90 million years.

Supercontinents are not built to last, however, and soon Pangea began to split apart. Burial of carbon subsequently decreased, and atmospheric CO_2 soon began to rise. A new greenhouse began around 250 million years ago, approximately coincident with the rise of the dinosaurs. Temperatures fluctuated during this interval, but were generally warmer than present. Eventually carbon burial began to increase yet again, as simple marine plants called phytoplankton increased in abundance in the newly expanded shallow seas. By 100 million years ago this trend reversed itself, and atmospheric carbon dioxide began to decline again. Hastening the process of global cooling was the growth of the Eurasian continent, which culminated with collisions between India and Asia, and between Saudi Arabia and Iran. Today Eurasia is girdled by a continuous chain of mountains from the Pyrenees and

Alps in the west, to the Himalayas in the east. A new ice age began in Antarctica about 34 million years ago, and Northern Hemisphere glaciers began to advance and retreat from centers in northern Canada, Scandinavia, and Siberia 2–3 million years ago. This ice age continues to the present day, although we have been enjoying a relatively warm reprieve for the past fifteen thousand years or so.

In absolute terms, the magnitude of human-induced climate change is puny in comparison to the grand fluctuations of the past. However, what anthropogenic climate change lacks in size and strength, it has made up in speed and agility. According to the 2013 report of the Intergovernmental Panel on Climate Change, 9.5 billion tons of carbon were released by fossil fuel combustion and cement production in 2011. This rate is more than one hundred times faster than the fastest long-term burial rates of organic carbon burial deduced from the geologic record (see Figure 2.7). We are therefore exhuming carbon from the Earth's crust much more quickly than it was originally buried.

Another way to look at this comparison is to examine past episodes of unusually rapid global warming. One of the most abrupt and best-known episodes occurred about 56 million years ago and has been named the Paleocene-Eocene Thermal Maximum (PETM). As with modern climate change, the PETM has been linked to a sudden increase in atmospheric CO_2 concentration. At the onset of this event, average atmospheric temperatures increased by approximately 5°C over a time frame of roughly ten thousand to twenty thousand years. Although extreme in geologic terms, a 5°C rise over 10,000 years works out to only 0.05°C per century. In contrast, average atmospheric temperatures have increased by about 1°C over the most recent century, a rate twenty times faster than that of the PETM.

Conclusions

Although the Earth might not be considered to be alive in a traditional sense, it does exhibit many of the attributes commonly used to define life, provided that you observe it over sufficiently long timescales. It has an active internal metabolism, its oceanic crust is continuously renewed, its continents grow through time, and its water and atmospheric gases respire in and out of the mantle. The Earth also responds to external stimuli, as applied chiefly by biological organisms living on its surface. On this living Earth, most resources are renewable. Virtually all of the known reserves of coal, oil, and natural gas have accumulated during the past 541 million years, which represent only about one-ninth of the total history of the Earth. If the Earth were a person, it would be near middle age; even if all the Earth's fossil fuel or other natural resources were expended, there is plenty of time left to accrue more. The time frames required for such planetary renewal stretch far longer than we can afford to wait, however.

For More Information

Berner, R. A., 2003, The long-term carbon cycles, fossil fuels and atmospheric composition: *Nature*, v. 426, p. 323–326.

Engels, Donald, Autumn 1984, "The length of Eratosthenes' stade": *American Journal of Philology* 106 (3): 298–311.

Gradstein, F. M, Ogg, J. G., Schmitz, M. D., et al., 2012, *The Geologic Time Scale 2012*: Boston, Elsevier, DOI: 10.1016/B978-0-444-59425-9.00004-4

Grotzinger, J., and Jordan, T. H., 2010, *Understanding Earth*, 6th ed.: W. H. Freeman, 672 p.

IPCC, 2013, *Climate Change 2013: The Physical Science Basis. Contribution of Working Group I to the Fifth Assessment Report of the Intergovernmental Panel on Climate Change* [Stocker, T. F., D. Qin, G.-K. Plattner, M. Tignor, S. K. Allen, J. Boschung, A. Nauels, Y. Xia, V. Bex, and P. M. Midgley (eds.)]. Cambridge University Press, Cambridge and New York, 1535 p.

Laske, G., and Masters, G., 1997, A global digital map of sediment thickness: EOS. Transactions, *American Geophysical Union*, v. 78, p. F483.

Peters, S. E., 2008, Environmental determinants of extinction selectivity in the fossil record: *Nature*, v. 454, p. 626-629.

Sleep, N. H., Zahnle, K., and Neuhoff. P. S., 2001, Initiation of clement surface conditions on the earliest earth: *Proceedings of the National Academy of Science*, v. 98, p. 3666–3672.

Valley, J. W., 2005, A cool early earth?: *Scientific American*, v. 293, p. 58–65.

3

Warmed from Above: Solar Energy

> At rest, however, in the middle of everything is the sun.
>
> *Nicolaus Copernicus, De revolutionibus orbium*
> *coelestium, 1543*

Our local star has continuously bathed the Earth in sunshine for the past four and a half billion years, interrupted only by the occasional solar eclipse. The Sun is expected to continue shining for billions of years to come, setting a gold standard of sustainability that no other major energy source can hope to match. The Sun supplies the energy for all plant life, and therefore is also the source of energy contained in both biofuels and fossil fuels. The Sun also powers the weather, through solar heating of the ocean and land surfaces. As will be discussed in the next chapter, hydroelectric and wind power are really just cleverly disguised versions of solar power. In fact, only two presently used energy sources can claim true independence from the Sun: nuclear fission and tidal energy. Nuclear fission relies on uranium, an element forged in the violent explosions of other, more massive stars called supernovae. Tidal energy results from the gravitational pull exerted on the ocean by the Moon, and to a lesser extent the Sun.

As powerful as it is, one thing the Sun cannot do on its own is to boil water at the ambient conditions found on the Earth's surface. At high noon on a clear day at the equator, sunlight heats the Earth's surface at a rate that is roughly equivalent to a portable electric hair dryer aimed at an average-sized desktop. The desktop (or ground surface) becomes warm, but not painfully so. At all other locations on the Earth the average rate of incoming solar power is less, because of clouds or to a Sun angle that is less than 90° relative to the horizon. Nowhere on Earth is it ever really hot enough to fry an egg on the sidewalk, a fact that can easily be demonstrated through independent study. It's a good thing, too, because otherwise the oceans would boil or evaporate, and the entire planet would be permanently cloaked in dense clouds. Viewed from the Moon, the Earth would look like a white pearl rather than a blue marble.

The constant rotisserielike movement of the Earth around its axis presents a major inconvenience for solar power production, because half of the globe is always kept in the dark. Some type of energy storage system will therefore always be required to take advantage of sunshine, after the Sun stops shining. On the other hand, the huge surface area of the Earth means that the total amount of solar energy striking it is immense, more than ten thousand times our current global power consumption. Of course, most of this sunlight total arrives in inaccessible locations such as oceans and mountainous areas, or else at high north or south latitudes, where it is too spread-out over the Earth's surface to generate much useful power. Even with these limitations, however, no other practical energy resource approaches the magnitude of solar. Additionally, solar power produces no greenhouse gases or other negative environmental effects, at least in principle. Its potential is therefore impossible to ignore.

Hell for Hydrogen: Nuclear Fusion in the Sun

It is commonly stated that solar energy is renewable, but this is not true in the very strictest sense. The Sun can be thought of as a sort of giant furnace that "burns" (figuratively speaking) atoms of the lightest chemical element, hydrogen. The Sun works with a finite supply of hydrogen, which was apportioned when it first formed. Eventually that supply will begin to run out, at which point the Sun is expected to bloat up into a red giant star that engulfs the Earth, and then finally collapse to a dense white dwarf star. Renewal or replenishment beyond that point is not in the cards; the best that can be hoped for is a sort of dim perdition. We know all this from observation of other similar stars that have already advanced into their old age. Fortunately, our own Sun's death throes are not expected to occur for another 5 billion years or so. Planet Earth should perhaps be concerned about the Sun's eventual extinction, but we need not be.

Hydrogen is the lightest chemical element, containing a single proton (positive charge) that constitutes nearly all of its mass. A single lone electron (negative charge) buzzes chaotically nearby. Hydrogen is by far the most common chemical element in the universe, making up an estimated three-quarters of its known chemical mass. The Sun and other stars hold huge amounts of hydrogen, but they don't literally burn it. Burning requires only the sharing of electrons with oxygen, a mundane chemical transaction that occurs every day all around us and within our own bodies. Hydrogen can in fact burn quite vigorously, as immortalized by the infamous explosion of the airship *Hindenburg* at Lakehurst, New Jersey in 1937.

A more fundamental process occurs within the Sun however: Hydrogen nuclei combine with each other to form new elements with larger nuclei, a process known as nuclear fusion. Four separate hydrogen atoms, with one proton each, create one helium atom with two protons and two neutrons. The electrons are more or less just along for the ride. About 0.7 percent of the original mass of the hydrogen is lost in the process of becoming helium; according to Einstein's famous equation

$$E = mc^2$$
(energy released = mass lost multiplied by
the speed of light squared)

The "c^2" part of this equation is a very large number, the speed of light, multiplied by itself. A small loss of mass therefore corresponds to an exceedingly large release of energy.

Hydrogen fusion doesn't happen just anywhere; it requires unimaginably high pressures and temperatures. This is true because the protons that constitute hydrogen nuclei carry a positive electrical charge, and like charges repel each other. The nuclei of heavier elements must therefore have another, stronger force, called the nuclear force, that counteracts electrostatic repulsion and binds protons together. The nuclear force only works when the protons are in intimate contact with each other, however. To get to that point they must first overcome the electrostatic repulsion that is felt even when they are located relatively far apart. You might think of this as being analogous to a deep well located on a tall mountaintop. If you wanted to drop a boulder into the well you would first have to roll it up the side of the mountain, at a considerable expenditure of energy. Once the boulder gets to the edge of the well, however, it will readily fall to the bottom and stay there.

The Sun is about 110 times the diameter of the Earth and has about 330,000 times its mass (Figure 3.1). Pressures in the Sun's core reach approximately 250 billion times the atmospheric pressure at

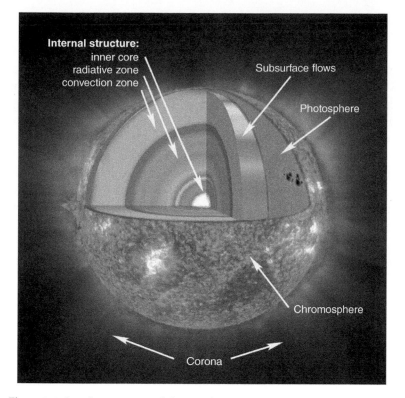

Figure 3.1. Interior structure of the Sun (NASA).

Earth's surface, and temperatures reach about 13.6 million K (K indicates the Kelvin scale, which has the same magnitude as degrees
Celsius but begins at absolute zero). These conditions provide the
energy in effect to "roll the boulder up the mountain," by crowding
protons close together and causing them to move at high rates of
speed that ensure they will collide with each other. Once they do, they
can fall down the nuclear "well," releasing tremendous amounts of
energy in the process.

The energy generated by fusion of hydrogen in the Sun's core
eventually works its way to its outer layers; by the time it reaches
the surface the temperature has dropped to a relatively balmy 5800K.
From there solar energy is radiated outward at a rate of 63 million
watts per square meter. To put this in perspective, a typical hair dryer
uses something in the range of 1,000 watts, so the intensity of solar
radiation is initially the equivalent of 63,000 hair dryers per square
meter of solar surface.

A Pleasant Warm Glow

Fortunately for us, the sunlight that actually reaches Earth is better suited to tanning than to instant incineration. A simple calculation shows why. The Sun is roughly a sphere with a radius of about 700,000 km. The energy leaving its surface radiates in all directions, spreading out as it travels toward the planets. The Earth is about 150 million kilometers away. At that distance, the Sun's energy can be thought of as illuminating the inside of an imaginary sphere with a surface area that is about 46,200 times greater than that of the Sun itself (Figure 3.2). The original 63 million watts per square meter emanating from the Sun is therefore reduced by the same factor, so that only about 1,367 watts per square meter arrive at the outer reaches of Earth's atmosphere.

The energy that actually penetrates the atmosphere and makes it to the Earth's surface is less, about 1,000 watts (a single hair dryer)

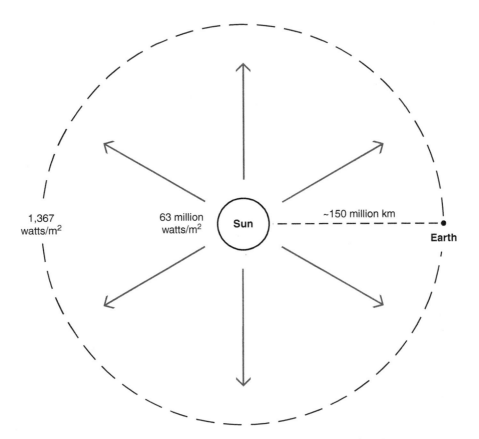

Figure 3.2. Illustration of the decrease in solar intensity with distance.

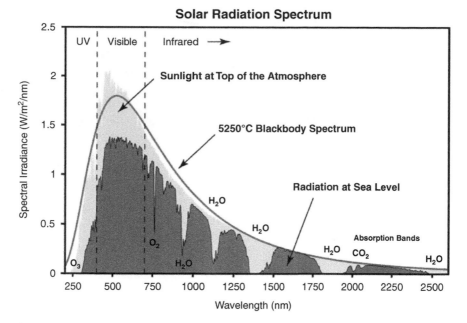

Figure 3.3. Solar radiation spectrum and atmospheric absorption (NASA).

per square meter at the equator, assuming a cloudless sky. Although the Earth's atmosphere appears transparent, it selectively blocks certain wavelengths of light, via absorption by water and carbon dioxide molecules (Figure 3.3). Ultraviolet light, which is responsible for sunburns, skin cancer, and the tacky "black light" posters that were popular during the 1970s, is largely blocked by ozone absorption (to be fair, ultraviolet light also allows our skin to produce vitamin D). Some wavelengths of infrared light, responsible for the invisible transfer of heat, are partially absorbed by water and carbon dioxide, as described in the previous chapter. In addition to absorption, air molecules cause scattering of incident sunlight in all directions. Short wavelengths, corresponding to blue colors, experience the most scattering. This explains why the sky looks blue and the sun looks yellow. Without atmospheric scattering the sky would look black and the sun white, just as it does from the atmosphere-free surface of the Moon.

Because the Earth is roughly spherical (technically an oblate spheroid), even tanning becomes a challenge at its higher northern and southern latitudes, a fact well known to pallid Alaskans. Neglecting the tilt of the Earth's axis, sunlight strikes perpendicular to its surface

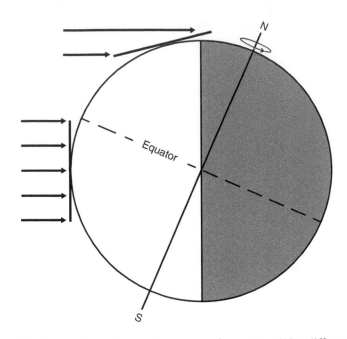

Figure 3.4. Geometric control on the amount of sunlight striking different parts of the Earth.
Parallel lines indicating incoming sunlight; straight lines tangential to the Earth indicate equal surface areas.

at the equator, and parallel to its surface at the poles (Figure 3.4). The total solar energy available at high latitudes should therefore be much less than at the equator, if expressed as watts per square meter of land surface. Scientists refer to the latter quantity as *insolation*, which is not to be confused with the more commonly used word insulation. Greater insolation implies less need for insulation, assuming you want to stay warm. To a first approximation, average annual insolation is highest near the equator and lowest near the poles.

In reality the Earth's axis of rotation tilts drunkenly at an angle of 23.5°, relative to a line perpendicular to the plane of its orbit. Its tilted axis also wobbles like a spinning top (you'd have to wait about twenty-three thousand years to witness one complete wobble). Axial tilt introduces a strong seasonal variation in solar isolation, most noticeable at higher latitudes. Sunlight-deprived residents of Alaska, northern Canada, northern Scandanavia, and Siberia know that the winter Sun barely clears the horizon, and north of the Arctic Circle fails to perform even that lazy feat during part of the year. Insolation therefore drops to zero. In contrast, the summers are warm enough to draw out the black flies and mosquitoes that often seem to constitute the principal population of those regions. Days are long, and for a time the Sun never sets north of the Arctic Circle. Peak summer insolation reaches its highest values at the poles, even though the Sun never climbs directly overhead. The same general pattern of seasonality prevails at the middle latitudes, where much of world's population lives, although the effects are less extreme.

Solar energy may seem to be a fair-weather friend; it is most available where and when it is least needed, at low latitudes and during the summer. However, the geometric limitations of low sun angle at high latitudes and during the winter can be largely corrected, by tilting a flat-plate collector upward so that the Sun's rays strike it perpendicularly. Regional insolation remains unchanged by this trick, but the energy output of the collector itself improves markedly. Optimal solar collector positioning can be a bit complicated to achieve, however. To a first approximation the collector should be tilted up at an angle equal to the local latitude. However, this angle fails to account for the changing seasonal trajectory of the Sun relative to the horizon. The collector should therefore be more steeply tilted in winter, and less so in summer. One could continually move the collector with the seasons, but depending on the installation this may not be feasible. If so, its fixed position must be optimized for its intended use.

Changes to incident sun angle due to the Earth's daily rotation are also problematic. To correct this problem, you could dash outside every hour or so to move the collector, or alternatively invest in a system of clocks and motors to do the job for you. In most cases, though, it is probably better just to leave the collector alone and accept that life isn't perfect. As it turns out, a well-positioned stationary collector can pull in around 70 percent of the energy that would be obtained by continuously tracking the Sun. The loss of energy caused by poor collector geometry during the early morning or late afternoon is less than might be expected, because the Sun's rays must penetrate farther through the atmosphere at those times. This in turn results in increased absorption and scattering, which reduce the intensity of the incoming light by about half when the Sun is near the horizon. Because less energy is available even to an optimally positioned solar collector, there is less energy to lose from an imperfectly positioned one.

Earth's Cloudy Veil

Clouds are the natural enemy of solar energy; they can reduce the total sunlight reaching the ground by factors of up to 80 to 90 percent. Fortunately our eyes and brains adapt automatically to these decreased levels of illumination; we may find cloudy conditions a bit depressing but we don't stumble around blindly for lack of light. Solar collectors, on the other hand, are immune to emotional disturbance, but have no choice but to downscale their energy output on cloudy days. The total amount of solar energy that is available at any given geographic location therefore depends in part on the local climate. Other factors being equal, Sun-scorched deserts will always out-perform overcast greenswards in terms of solar energy potential.

Ironically, the Sun itself is directly responsible for the existence of clouds. Incoming sunlight mostly strikes the ocean, since the ocean accounts for 71 percent of the surface area of the Earth. This in turn results in partial evaporation of surface waters, which increases the moisture content of the overlying air. The warm air mass rises to elevations where lower atmospheric pressures prevail, causing it to expand and cool. The same basic phenomena can be duplicated by releasing air from a tire, or from a compressed air hose. In both cases the valve or discharge nozzle will become colder.

Cold air is less energetic than warm air and therefore can hold less moisture. Rising air masses may eventually reach a limit called the dew point temperature, at which point water begins to condense out of the air to form the small droplets that constitute clouds. Put differently, the dew point is the temperature at which the air is 100 percent saturated with water and can hold no more. Airplane pilots can use this information to estimate the height to cloud bases; they need only know the surface temperature, the dew point temperature, and the rate at which temperature decreases with altitude (which is approximately linear over the altitude range of interest). From this information they can deduce the altitude at which 100 percent saturation will occur.

For clouds to obscure the continents, winds must blow moist air onshore. The Sun's meddling is once again at play, this time in the form of uneven heating of the Earth's surface. During the summer months, land surfaces tend to heat up dramatically, often increasing in temperature by tens of degrees Celsius. In contrast the ocean plays it cool, changing temperature by a few degrees at most. Warm air therefore tends to rise above the warming continents, and is replaced by inflowing cool, moist air from the surrounding ocean. This damp marine air is ultimately responsible for the creation of clouds. During the winter months the situation is reversed; the continents lose their heat just as fast as they gained it, becoming *colder* than the ocean. Descending air and higher pressures prevail on the continents, resulting in offshore winds and winter dryness. Sales of sweaters and skin lotion go up in tandem.

If clouds are the enemy of solar power, then deserts are its friend. On a year-round basis, deserts commonly receive twice as much sunlight as do more hospitable regions at similar latitude. This is not to say that useful solar power cannot be generated in cloudy regions; certainly it can. However, the same investment in solar collection systems would buy up to twice as much power output if they were installed in the desert (other factors being equal). This basic physical reality is hard to ignore and leads to the conclusion that solar power is to a large extent a geographically heterogeneous resource.

The Global Geography of Solar Energy

It is probably no accident that the English word "desert" can be used as a noun indicating a barren, desolate, or dry place, and also as a verb meaning to leave or abandon. The term in its former sense is commonly applied to areas with less than 25 cm/yr of precipitation; a more complete definition specifies that the potential for evaporation exceeds annual rainfall. Humanity has for the most part avoided such regions, preferring to leave them to lizards, nomads, fortune seekers, and persons prone to experiencing mystical visions. Lack of water may perhaps be responsible for some of those visions, and it strongly inhibits the development of major population centers.

Depending on how you define them, deserts represent something like one-third of the Earth's land surface area. Why does the Earth have so much apparently worthless real estate? The answer to this question is multifold, but ultimately tied to the geometry of the Earth, its ocean, and its continents. To a first approximation deserts are planetary-scale features, related to the roughly spherical shape of the Earth and to the physical properties of its atmosphere.

The story begins at the equator. Equatorial regions on average receive more sunlight per square meter than do polar regions, resulting in a large geographic temperature gradient. The atmosphere, on the other hand, is a fluid that wants to erase this gradient, by carrying heat toward the poles. Warm equatorial land surfaces heat up the

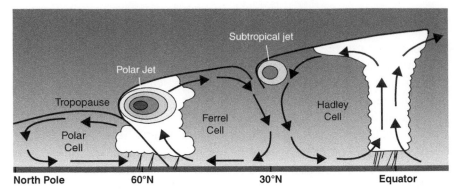

Figure 3.5. Generalized illustration of global atmospheric circulation (NASA).

overlying air, causing it to rise in much the same way as a hot air balloon (Figure 3.5). As it rises it cools, condensing its moisture into rain that drenches the Earth's surface. The clouds associated with rising moist air tend to reduce the amount of sunlight reaching the surface in a narrow band near the equator (Figure 3.6).

The air generally stops rising when it bumps into the overlying stratosphere, where temperatures cease to decrease and then begin to increase with increasing altitude. Since rising air masses no longer continue to go up, they instead spread north and south away from the equator. They don't make it all the way to the poles, however, because of air mass cooling and the rotation of the Earth. The former tends to make the air sink toward the surface again, and the latter causes the Coriolis effect, which bends large-scale air currents to the right in the northern hemisphere, and to the left in the southern hemisphere.

In both hemispheres the originally poleward flow of air turns toward the east because of Coriolis deflection, and descends back to the surface at latitudes of roughly 30° north and south (known to sailors as the "horse latitudes"). The air is repressurized as it descends, causing it to warm up. This process is analogous to inflating a tire, causing the tire valve to heat up. Because warm air can absorb more moisture, descending air inhibits the formation of clouds, resulting in clear skies and widespread formation of deserts (Figure 3.7). The major desert tracts of North Africa, Central Asia, and Australia are attributable in part to this phenomena. The low-latitude atmospheric loop is closed by surface flow toward the southwest or northwest, back toward the equator, known as the trade winds.

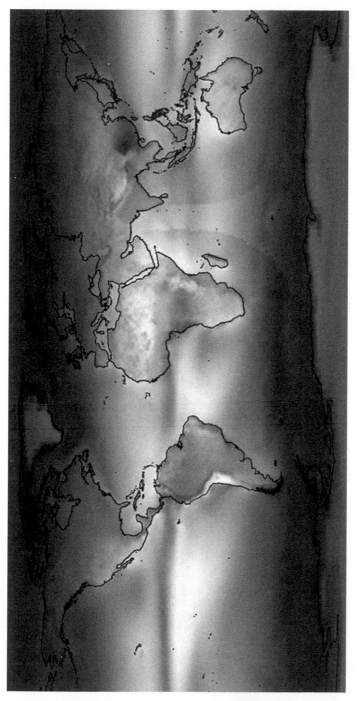

Figure 3.6. Average insolation across the Earth's surface, based on satellite measurement. Lighter areas receive more total solar radiation (NASA Surface Meteorology and Solar Energy Project).

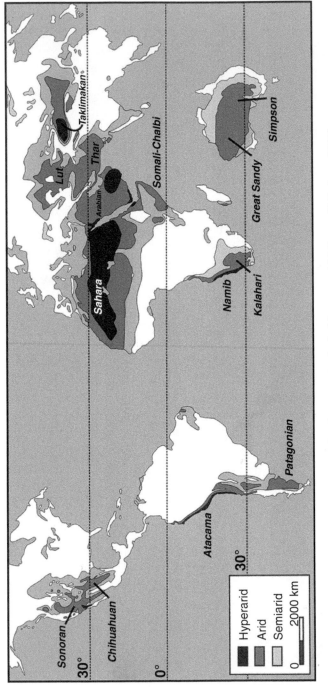

Figure 3.7. Distribution of world's deserts (modified from U.S. Geological Survey and Maliva and Missimer, 2012).

Silk Roads and Swirling Seas

The size and topography of the continents further shape the distribution of the world's deserts. At the simplest level, this occurs because moist air masses moving from the ocean onto the continents lose some of that moisture whenever it rains or snows. The air therefore becomes progressively drier the farther it penetrates inland, making the interiors of continents drier on average than their edges. Asia is the largest continent, and the aridity of its core has historically isolated China and other eastern Asian countries from Europe and the Middle East. Marco Polo and others who journeyed between the two regions did so via the "Silk Roads," which in fact consist of a series of oases that dot the Central Asian desert.

Peripheral mountain ranges can make continental interiors even drier, by forming a barrier to the flow of moisture inland. As moist air moves toward the mountains it is forced upward, causing it to cool and lose water as precipitation. The most dramatic example of this process is seen during the Asian monsoon, which occurs during the summer months when moist air flows from the Indian and western Pacific Oceans toward the warm interior of Asia. Tremendous rainfall occurs on the windward face of the Himalayas and other ranges, providing the water supply for approximately one-third of the world's population. The lee sides of the ranges are proportionately drier, however, and experience a rain shadow effect as the moisture-depleted air moves inland and descends off the mountains.

Although the largest deserts lie within continents, some of the most extreme deserts lie at their edges. Ironically the world's driest desert, the Atacama in northern Chile and adjoining areas, is located immediately adjacent to the world's largest body of water, the Pacific Ocean. The Atacama lies within dry horse latitudes of about 20°S–30°S, and the adjacent Andes Mountains impose a rain shadow effect on northwest-flowing Atlantic moisture. Both of these factors contribute to the aridity of the Atacama desert, and the coup de grâce is delivered by the Peru Current, an ocean current that flows northward and parallel to the coastline. Coriolis deflection of the Peru Current redirects surface waters westward, causing them to flow away from the Chilean coast. This offshore flow is simultaneously balanced by coastal upwelling of deeper, colder waters. The overlying coastal air is in turn chilled by these cold waters, which dry it out and reduce its ability to carry moisture onshore. Extreme aridity related to coastal upwelling is a common feature on many other west-facing coastlines as well, for example, in Baja California (Mexico), coastal Morocco, and southwestern Africa.

Sun, Sand, and Silicon

As noted earlier, much of the world's population lives in cloudy reaches of the Earth that experience relatively modest levels of solar insolation. The wealthier inhabitants of these dampened regions often seek a respite from the gloom when they go on vacation, commonly traveling to sunnier locales. As they soak up the Sun's rays and sip overpriced tropical drinks, a few may even gaze lazily upward to contemplate the bounty of solar energy arriving from above and its promise for planetary salvation. They probably do not realize, however, that the raw material for most presently used solar technology lies directly beneath their feet, in the form of beach sand. Sand is used to make glass, which has some uniquely useful properties for building solar energy collectors.

Solar heating of water, air, or other fluids relies on a remarkable property of glass: It allows visible light to pass through virtually unimpeded, but is opaque to long-wavelength infrared light (= heat). Greenhouses have long exploited this principle to allow year-round cultivation of plants in cold climates, as explained in the previous chapter. Solar thermal collectors are similar in design to greenhouses, except that the accumulated heat is transferred away from the collector for other uses (such as heating a bath). Even ordinary south-facing windows serve much the same function (in the Northern Hemisphere), by allowing the sun to heat interior air. This is a truly ancient technology; the Romans knew how to make window glass and may have used it to warm the interior of their bathhouses.

Sand consists of nothing more than broken bits of rock, gathered together and sorted by river currents, wind, waves. Viewed under a microscope sand is a rock-hound's collection in miniature, with surprisingly diverse colors, shapes, and textures. Most sand contains the mineral quartz, an unusually hard and chemically stable mineral. Quartz sometimes takes the form of spectacular six-sided crystals, believed by some to be imbued with mystical powers. Quartz is built from oxygen and silicon, the #1 and #2 most abundant chemical elements in the Earth's crustal top 20. Other minerals in sand likewise contain oxygen and silicon, but also reach further down the list to include elements such as aluminum (#3), iron (#4), and others.

Glass contains the same basic chemical constituents as quartz, but lacks its orderly crystalline structure. To understand this distinction consider a football stadium filled with well-behaved spectators, sitting in carefully arranged rows of assigned seats of identical size that are fixed at a specific distance from each other. This arrangement of people is analogous to the regular arrangement of atoms in a quartz crystal. Now, imagine that at the end of the football game the supporters of the victorious team flood onto the field in zealous frenzy. The previously well-behaved spectators now become simple hooligans, shouting, spilling beer, and running about in crazy circles. Glass can be thought of as a collection of atomic-scale hooligans; it contains the same atoms as quartz but they are definitely not well behaved. Glass is made by melting quartz at temperatures in excess of 1,600°C, which wipe out its original crystalline structure. It is then quickly quenched at a lower temperature, causing it to solidify before the silicon and oxygen atoms can find their way back into their assigned seats.

Photovoltaic cells do not require the use of glass in their manufacture, but they still do rely on geologic supplies of silicon. Their crystalline structure is revealed to the casual observer by the way that such photocells reflect more light when held at certain angles. Unlike quartz, the silicon crystals used in photocells do no occur naturally; they must be manufactured. Ironically this means that most photoelectric cells consume a substantial amount of energy input before they can produce any energy output, because of the high temperatures needed to melt sand, separate silicon, and grow silicon crystals.

To separate silicon (chemical symbol Si) from quartz (chemical formula SiO_2), molten quartz is reacted with carbon at temperatures approaching 2000°C, equal to about one-third of the temperature at the Sun's surface. Silicon crystals then must be formed slowly from a melt that is maintained at about 1,500°C. The most efficient photovoltaic cells in common use are made from monocrystalline silica, which ultimately requires energy inputs approximately equal to 10 percent of the energy that is generated over the lifetime of the cell. This up-front energy investment translates into substantial capital cost for installed solar cells. However, the patient will be rewarded with many years of low-cost electricity production.

Not just any sand will do when it comes to making high-quality glass or silicon crystals; ideally it should be as near to pure quartz as possible. Such deposits are actually somewhat unusual geologically; they result from natural purification processes that acted over millions of years. For example, some of the best glass sands in the United

States were deposited about 450 million years ago, when shallow seas washed across much of the continent. The relentless pounding of sand by waves, combined with sandblasting of dune sands by wind, gradually destroyed most grains weaker than quartz. A corrosive atmosphere with carbon dioxide levels ten times that of today helped to dissolve whatever minerals survived the beating, save impervious quartz. Fortunately for us, the wide geographic distribution of the ancient seas helped to ensure a plentiful supply of quartz sand. There is virtually no chance of shortage in the foreseeable future.

(Un)Moved by the Sun

Even in the driest of low-latitude deserts solar energy is a gentle giant: immense in size, but never overly assertive. The diffuse disposition of sunlight imposes some intrinsic limitations on its practical use as an energy source, particularly with respect to converting solar energy into motion.

Steam offers a simple and inexpensive means of converting thermal energy into mechanical energy, taking advantage of the volume expansion that occurs when water is heated and vaporized. We tend to think of steam engines as outdated relics of our primitive past, largely of interest to museums and certain railroad enthusiasts, but nothing could be further from the truth. Most modern electrical power plants use steam to drive turbines, which in turn drive generators. Today's cutting-edge computers and the Internet are for the most part powered by steam!

Steam turbines and all other heat engines are subject to certain inescapable physical limits on the efficiency with which they convert heat into motion. The thermal efficiency of an engine can be defined as the percentage of input heat energy that can be used to do useful work. Early in the 19th century the French physicist Sadi Carnot noted that the maximum possible efficiency of a heat engine is strictly governed by the temperature difference between an engine's hot operating fluid (burning gases, steam, or just hot air) and the cooler ambient temperature of its surroundings. In the case of a typical coal-fired electrical power plant, the operating fluid is pressurized steam or supercritical water, heated to temperatures near 400°C. The temperature of the cooling medium (air or water) might average around 15°C. This temperature spread gives a maximum efficiency, under ideal conditions, of around 54 percent. However, this ideal or "Carnot" efficiency really only applies to machines that are moving at an infinitesimally slow speed and fails to take into account various other unpreventable energy losses. In practice the electrical energy output of a typical coal-fired power plant amounts to only 30 to 35 percent of the heat provided by combustion.

Raw, unfocused sunlight at the Earth's surface cannot approach the operating temperatures of a coal-fired power plant. A more realistic point of comparison might be a minivan with a burgundy interior, parked at a suburban shopping mall near Phoenix in July, at lunchtime. Lets assume the car's owner has left a small dog named Fifi sitting inside to repel any would-be intruders and closed all the windows. Fifi may ultimately expire from dehydration, but the dog will never burst into flame. The maximum temperature inside the car might reach 60°C, but will not exceed the melting point of the scorching vinyl seats (about 85°C). The temperature difference between the inside of the car and dry heat outside would therefore be only about 20°C–25°C. The *maximum* thermal efficiency calculated from this temperature difference is only about 8 percent, and the actual efficiency of a solar heat engine installed in the car would be much less than that. The car's hot interior would therefore be virtually useless for generating any appreciable motive force.

An alternative approach to solar transportation is to use photovoltaic cells to convert sunlight directly into motion, avoiding heat engines and their limitations altogether. Photovoltaics work by using sunlight to force electrons to move between two adjacent semiconducting materials, which most commonly are constructed by coating crystalline silicon with elements such as phosphorus and boron. The moving electrons can be used to operate an electric motor, which in turn can propel an automobile.

This idea looks great on the drawing board; after all, what's not to like about a car that needs no fuel and produces no emissions? Regular competitions are held between teams of university engineering students to produce just such a vehicle (the Solar Challenge), and these inspired students have produced many impressive designs, capable of carrying a single driver at reasonable speeds down a smooth road on a sunny day. Such a car would not satisfy the practical demands of most motorists, though, particularly those who prefer large sport-utility vehicles (SUVs). Some solar car designs bear a curious resemblance to a rolling slice of toast, with lots of flat upper surface area covered in photovoltaic cells. The driver's head often protrudes beneath a clear plastic bubble, like a dollop of jam on top of the toast.

The reason for this cramped seating is simple: There just isn't enough solar energy available to obtain the performance of an average car, much less an SUV. The mild nature of sunlight is once again to blame. A typical smallish car might have a surface of about 8 square meters, which on an ideal day in the Sahara desert might receive sunlight at the rate of about 1,000 watts per square meter, for a total of 8,000 watts. Converting units, this is equal to only about 11 horsepower (hp). As it happens, an actual horse can produce this much power for brief periods. The sporty Toyota Prius hybrid has 80 horsepower on tap, and larger SUVs usually have at least 200. With some major scaling back of our expectations a practical 11 hp car might be possible, but at this point efficiency once again raises its ugly head. The actual electricity produced by photovoltaic cells mounted on a solar car will be only around 10 percent of the available solar energy. So rather than 11 horsepower you have about 1 horsepower at best, and most of the time you will have less. Perhaps it would make more sense to buy a living horse instead, learn to ride, and relive the halcyon days of the 18th and 19th centuries!

Practical Use of Solar Energy

In terms of the preceding considerations its no surprise that sunlight does not power planes, trains, or automobiles; its hotter hydrocarbon cousins are simply better suited to such jobs. For sunlight to be useful as an energy resource we either need to concentrate it beyond its natural intensity or else to find jobs that are better suited to low-intensity energy. The first strategy is familiar to anyone who has used a magnifying glass to burn paper or to incinerate ants. The magnifying glass focuses a relatively large area of sunlight, equal to the surface area of the glass, down to a much smaller spot with much higher temperature. The same principle can be scaled up for commercial power generation, by using mirrors or lenses to concentrate the sunlight striking a relatively large land surface area onto a relatively small energy collection device, such a as boiler driving a steam turbine. Such systems must be able to track the Sun's path actively to remain in focus and are relatively expensive to construct and maintain. They therefore make the most sense when installed where the available solar resource is greatest, in deserts. The world's largest concentrated solar collection system, the 392 megawatt Ivanpah Solar Electric Generating System, is currently being installed in the Mohave Desert in eastern California.

Although photovoltaics are a poor choice for powering cars, they do have their uses. The most obvious is to provide electricity in places where no other viable options exist. One of their earliest practical uses was to power orbiting spacecraft, and photovoltaics are commonly used in isolated terrestrial regions that are not connected to a power grid. Photovoltaics have also become increasingly popular for residential installation in urban areas, on the principle that what they lack in efficiency they can make up in size. For example, in many regions a house with a south-facing rooftop can generate enough electricity to meet the needs of its residents, and during peak hours it may actually be possible to sell electricity back to the local utility. The same utility, or a system of storage batteries, is needed to provide power after the sun goes down. The main barrier to such applications is high initial acquisition cost, but a patient homeowner may eventually break even. As the technology of solar photovoltaic cells continues to evolve, hopefully their cost will continue to decline; a multitude of new technologies are being investigated and laboratory efficiencies approaching or even exceeding 40 percent have been reported. Such high efficiencies remain in the realm of the experimental, however; at present solar photovoltaics are the most expensive way to generate electricity.

Sunlight really comes into its own when tasked with producing relatively small increases in the temperature of air or water. In fact its main work on Earth is to do just that, by driving the planetary-scale circulation of the atmosphere and ocean currents. Solar heating nicely matches the requirements of residential space and water heating. The temperature requirements for staying warm and clean are relatively modest: Furnaces and water heaters need only raise ambient air and water temperatures by 40°C–50°C at most. Efficiency in this case can be thought of as the proportion of incoming solar radiation that can be prevented from immediately escaping back into the atmosphere, via reradiation from hot surfaces, heat conduction through the solar collector itself, or heat convection by air currents. Whereas small temperature contrasts hinder the efficiency of heat engines, they are actually beneficial in slowing the transfer of heat away from heated interiors or hot water systems. According to Professor S. A. Klein, director of the University of Wisconsin-Madison Solar Energy Laboratory, a well-designed solar water heating system should be able to obtain 40 percent efficiency. Unlike futuristic solar cell technologies, solar water heating is available now.

Currently more than half of the energy used by an average residence in the United States goes to space heating and water heating, with natural gas being the most common source. Using natural gas for residential air and water heating is a lot like using a welder's torch to heat soup. The soup will quickly get hot, but one can't avoid a certain sense of comic overkill. After the novelty wears off the welder will want his torch back to do more serious work. Natural gas and other hydrocarbon fuels likewise have bigger jobs to do than heating air and water by a few tens of degrees. Given their limited abundance it would make sense to prioritize the use of such fuels for applications where they are most useful.

Star in a Bottle: Nuclear Fusion

The limitations of solar energy stem entirely from the great distance between the Earth and the Sun; sunlight unavoidably spreads out and becomes weaker as it travels farther from its source. If we could somehow move the Sun closer to the Earth the intensity of available solar energy could be dramatically increased, but unfortunately we would not survive the increased surface temperatures long enough to enjoy this newly boosted energy supply.

What if we could instead build a miniature version of the Sun on Earth? We could then gain the advantages of an efficient high temperature energy source, without roasting the planet in the process. Ongoing research on nuclear fusion aims to realize this goal, by eventually building practical reactors that mimic the Sun. The technical challenges to doing this are steep, however, in part because of the need to heat the starting materials to temperatures like those found at the center of the Sun. Since no solid material can withstand such temperatures, confinement of the fusion reaction must be accomplished by other means, such as strong magnetic fields. A large amount of power is required just to initiate the fusion reaction, and only recently have fusion experiments reached the threshold of yielding an equal amount of power back.

Rather than use hydrogen as the starting ingredient, most power-related research has focused on fusion of its heftier cousins deuterium and tritium. Deuterium is hydrogen with one neutron added, giving it an atomic mass of 2 (hence its name, derived from the Latin word for "two"). Tritium adds two neutrons, for an atomic mass of 3. Deuterium and tritium are both isotopes of hydrogen, meaning they have the same number of protons but different numbers of neutrons. Deuterium is readily available from seawater, where it naturally accounts for about 0.015 percent of the hydrogen found in water molecules (H_2O). Tritium, in contrast, is exceedingly rare in nature and must therefore be produced by bombarding lithium with neutrons in a nuclear reactor.

Various combinations of these and light isotopes could also be used in fusion reactors to produce power, but the deuterium-tritium reaction has received the most attention because it is the easiest to achieve. Deuterium-tritium fusion also has a significant downside, however; each reaction emits a high-energy neutron, and neutron absorption by the reactor walls gradually damages them and makes them radioactive. The wall materials must therefore be replaced every few years, and once removed constitute high-level radioactive waste that must be safely stored for at least fifty years.

An alternative pathway is to fuse deuterium with helium-3 (^{3}He), which contains one less neutron than common helium (^{4}He). This reaction emits a proton rather than a neutron and therefore produces far less radioactive waste. Another advantage is that the protons, being charged particles, can themselves be used to produce electricity directly, rather than requiring the intermediate step of boiling water to make steam. The theoretical efficiency of power production could thus be much higher than for deuterium-tritium fusion. Deuterium-^{3}He fusion does have some problems of its own; for example, the energy required to achieve this reaction is much higher than for deuterium-deuterium. Even if this problem can be resolved through clever reactor design, another major obstacle looms beyond it: Very little ^{3}He exists on Earth. Only about one in a million helium atoms on Earth consists of ^{3}He; present commercial supplies are limited to a few hundred kilograms derived mostly from the radioactive decay of tritium used in nuclear weapons.

Most of the naturally occurring ^{3}He in the solar system resides within the Sun, where it is impossible to access directly. However, the Sun ejects ^{3}He regularly as part of the solar wind, a fast-moving stream of charged particles traveling outward through the solar system. The density of ^{3}He in the solar wind is very low, but if you could set up some kind of collector in the path of this flow and keep it there for a sufficiently long time you could nonetheless capture a large supply. Unfortunately you could not build such a collector on Earth because the solar wind is deflected by the Earth's magnetic field. We get spectacular nocturnal light displays (aurora) as a result, but no ^{3}He to speak of.

The Moon, however, has a very weak magnetic field, owing to its lack of a liquid iron outer core like that of Earth (see Chapter 2). It also has no appreciable atmosphere that might interfere with incoming helium ions. It can therefore capture ^{3}He from the solar wind and has been doing so for billions of years. Initially these atoms are embedded only on the surfaces of exposed lunar rocks and dust, but ongoing meteorite impacts continually pulverize the Moon's surface into sand- and dust-sized particles. Over time these particles have built up into a kind of soil called lunar regolith, which is thought to reach thicknesses of up to 6 m in some areas. Continual churning or "gardening" of the regolith by meteorite impact mixes ^{3}He deeper beneath the lunar surface, thereby preserving substantial deposits.

Analysis of samples returned by NASA's Apollo missions confirms that the lunar regolith and associated rocks do contain ^{3}He, at concentrations typically in the range of 10–20 parts per billion by weight. While these concentrations may seem low the lunar regolith is relatively soft and easy to excavate, and its ^{3}He can be readily released by heating the deposits to temperatures in the range of 600°C–800°C. Proposals for mining lunar ^{3}He began in the mid-1980s, led by nuclear engineer Gerald Kulcinsky and colleagues at the University of Wisconsin-Madison. The total available lunar resource has been estimated at about 1 million tons, which would be the energy equivalent of roughly ten times the presently recoverable coal, oil, and gas on Earth. Fusion using lunar ^{3}He might therefore ensure a reliable and safe energy supply for the next one thousand years or so.

There is of course some inconvenience involved in getting to the Moon to mine this resource, but its extremely high energy density might well make the trip worthwhile. Former Apollo astronaut Harrison Schmitt, the only geologist ever to walk on the Moon, has long advocated such a plan, and presented a detailed analysis of the economic and other issues related to lunar mining in his 2005 book *Return to the Moon*. On the basis of prevailing fossil fuel energy prices he estimated the value of ^{3}He to be about $140 million per 100 kg, which works out to $40,000 per ounce. This would make it about twenty to forty times more valuable than gold! The size of the prize could thus be very large indeed and by Schmitt's analysis would be more than enough to repay the required investment. Sending spacecraft to the Moon to mine ^{3}He may seem farfetched and speculative, but perhaps it is no more so than some of the other major historical explorations that have occurred on Earth? Although clearly very ambitious in scope, mining the Moon may actually pose less of a scientific challenge than developing the commercial fusion reactors that would use the resultant fuel.

Conclusions

Solar energy embodies a fundamental paradox: The total quantity reaching the Earth is enormous, but the rate at which it arrives at any given location is rather modest. The availability of solar energy is also strictly governed by the rotation of the Earth, which requires either the pairing of solar with another energy source or the construction of new energy storage systems. Balanced against these limitations is the fact that solar power produces no intrinsic waste products, unless you count helium produced in the Sun (most of which remains there).

The diffuse nature of sunlight in its raw form makes it poorly suited to high-demand jobs, such as steam-operated electric power plants or automobile propulsion. Solar energy therefore requires some form of artificial concentration to do any really heavy lifting. This can be done in multiple ways, for example, by building large arrays of photovoltaic cells or by using mirrors to focus light from a large area onto a small one. Such technologies have long been used for electricity generation, but have traditionally carried higher costs than competing energy sources. These costs have been declining, however, and future improvements in solar technology hopefully will continue the trend.

Solar collectors cost about the same regardless of where they are installed, so it makes the most sense to build them where the Sun shines most consistently. To a first approximation the availability of solar power is tied to the geography of low- to midlatitude deserts, which can receive twice as much sunlight as more humid regions at similar latitudes. The Earth is well supplied with suitable desert regions, which in principle could easily satisfy all our present and future energy needs. The only flaw in this logic is that most of the world's population does not live in deserts, and that implies the need for some way to transport power to where they do live. The best use of solar energy, at least in the short term, may be for low-demand jobs such as water and space heating. The technology of such applications is relatively simple and well established, and the resultant reductions in the use of other "higher-quality" energy sources can be considerable.

For More Information

Berman, Robert, 2011, *The Sun's Heartbeat: And Other Stories from the Life of the Star That Powers Our Planet*: Back Bay Books, 320 p.

Daniels, F., 2010, *Direct Use of the Sun's Energy*: Ishi Press, 404 p.

Douglas, R. W., and Frank, S., 1972, *A History of Glassmaking*. Henley-on-Thames, G. T. Foulis, 213 p.

Goetzberger and Hoffman, 2005, *Photovoltaic solar energy generation*: Springer, 248 p.

Kallberg P., P. Berrisford, et al. (2005). ERA-40 Atlas: European Centre for Medium Range Weather Forecasts.

Maliva, R., and Missimer, T., 2012, *Arid Lands Water Evaluation and Management, Environmental Science and Engineering*: Berlin Heidelberg, Springer-Verlag, 1076 p.

Ring, J. W., 1996, Windows, baths, and solar energy in the Roman Empire: *American Journal of Archaeology*, v. 100, p. 717–724.

Schmitt, H. H., 2006, *Return to the Moon: Exploration, Enterprise, and Energy in the Human Settlement of Space*: Copernicus, 352 p.

Stevenson, R. D.; Wassersug, R. J. (1993). "Horsepower from a horse": *Nature* 364 (6434): 195.

Wittenberg, L. J., Santarius, J. F., and Kulcinski, G. L., 1986, Lunar source of ^{3}He for commercial fusion power: *Fusion Technology*, v. 10, p. 167–178.

Würfel, Peter, 2009, *Physics of Solar Cells:* Wiley.

4

Wind, Water, and Waves: Energy from the Fluid Earth

The wind goeth toward the south, and turneth about unto the north; it whirleth about continually, and the wind returneth again according to his circuits. All the rivers run into the sea; yet the sea is not full; unto the place from whence the rivers come, thither they return again.

Book of Ecclesiastes, King James version

If you don't like the weather around here, just wait 5 minutes and it will change.

Universally expressed local opinion about all weather, everywhere

Sunlight is abundant, free, and, from the human perspective, eternal. However, as noted in the previous chapter, sunlight is not particularly strong in its natural state, and requires artificial concentration if it is to do any really ambitious work. Fortunately when it comes to concentrating sunlight, nature has already taken the lead. This is true because sunlight causes dramatically unequal heating of the Earth's surface at different latitudes, between land and sea, and between night and day. Uneven heating sets air and water into motion, leading to a complex and sometimes violent sequence of events that act to spread warmth more evenly across the globe. These events never reach an end, however, because the Sun continues to shine and the Earth never stands still. We therefore experience perpetually dynamic weather.

Wind can thus be thought of as solar energy converted into motion. The interaction between wind currents and surface topography help to focus this flow, resulting in localized areas where surface wind speeds exceed the regional average. An opportunity therefore exists for the wind to do real work, if we are somewhat selective about where and when that work is to be done. This is of course no great revelation; wind has propelled sailing ships and turned windmills since the dawn of recorded history. However, wind has also developed a well-deserved reputation for inconstancy. Some days it arrives with a destructive vengeance that precludes any peaceable use; other days it fails to show up at all. Surface wind speed routinely increases during the daylight hours, but then drops off in the early evening, just as residential electric loads are ramping up. For wind to shoulder a large share of our primary energy load therefore requires that the impact of such lapses be ameliorated.

Wind also propels waves, by blowing over the surface of the ocean (or across a lake). Tides are also a kind of wave, which rises and falls over time frames of hours rather than seconds. Tides are unique, however, in that they derive their energy directly from gravitational interaction of the Earth, Moon, and Sun, with no need to involve wind as an intermediary. Energy may be recovered from all forms of water waves, in proportion to the height difference between their peaks and troughs and the time that elapses between successive waves. The gravitational potential energy of a single wave is relatively small, but what waves lack in stature they make up in endless repetition.

The great majority of Earth's wind-driven wave energy lies beyond easy reach, in the open ocean far from shore. Furthermore, the magnitude of tides in the open ocean is generally less than one meter. It may be technically possible to recover wave and tide energy from such distant seas, but it will likely require an elevated level of desperation before we actually attempt to do so. The situation is more encouraging near the seacoasts, however, in part because this is where much of the world's population lives. Natural variations in coastal geometry can also amplify the magnitude of waves and tides as they reach the shore, providing natural concentration of a normally diffuse energy resource.

At first glance hydroelectric power would seem to have little relation to the Sun, but this appearance is deceiving. Hydroelectric dams work by exploiting the difference in gravitational potential energy between an upstream reservoir and its lower elevation outflow. Reservoirs require rivers to fill them, and rivers require precipitation in the form of either rain or snow. Precipitation originates from water vapor contained in the atmosphere, put there by evaporation of ocean water by the Sun. Hydroelectric, therefore, is just another way of collecting solar energy, which relies on natural river watersheds to focus the gravitational energy contained in continental precipitation.

The Planetary Perspective on Moving Air

Wind is moving air, and air is a gas that comprises about 78 percent nitrogen, 21 percent oxygen, and 1 percent other gases (mostly argon). The French scientist and inventor Jacques Alexandre César Charles is credited with discovering in the late 18th century that all gases expand in direct proportion to their temperature when heated, regardless of their composition. Charles's law, as it has been known to subsequent generations of chemistry students, is the basis for converting sunlight into wind. Solar energy that warms the Earth's surface is reradiated as heat, which in turn warms and expands the overlying layer of air. The resultant decrease in density sets the air into motion, causing it to rise to higher levels in the atmosphere just as a hot air balloon would do.

As described in the previous chapter, the most intense solar baking occurs near the equator, triggering an atmospheric loop called a Hadley cell at low latitudes (Figure 3.5). At around 30° north and south the Hadley cells link with midlatitude Ferrel cells, which mirror the low-latitude circulation. Beyond the Ferrel cells lie polar cells, completing an atmospheric bucket brigade that continually tries to extinguish equatorial heat by spreading it out across the planet. Each of the circulation loops generates its own distinctive surface wind patterns: easterly trade winds below the Hadley cells, prevailing westerlies below the Ferrel cells, and polar easterlies below the polar cells (Figure 4.1). Surface winds tend to slacken between the cells, causing the "doldrums" at the equator and calm "horse latitudes" near 30° north and south.

Of course, if global wind patterns were really this simple then there would be no need for wind vanes or weather forecasters; conditions everywhere would be as tedious and unchanging as summertime in Los Angeles. Reality in most locales is more exciting, however, because of the tilt of the Earth's rotational axis and the haphazard scattering of continents across its surface. The former causes seasons. The latter is important because of the markedly different thermal properties of sea versus land; the intrinsic heat storage capacity of water is about twice that of rocks, and waves and currents continually stir the upper part of the ocean. The temperature of the ocean's surface therefore only changes by a few degrees between summer and winter, whereas the land surface changes by tens of degrees. Summertime

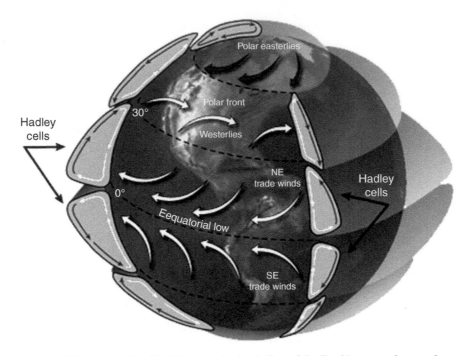

Figure 4.1. Simplified large-scale circulation of the Earth's atmosphere and surface wind patterns (NASA).

warming of the continents induces the air to rise above them; this rise is balanced by descending air over the oceans. In the winter the land cools more than the ocean, reversing these trends. Up-and-down movements of air at the continental scale relieve some of the monotony of planetary-scale convection.

At even finer scales, the boundaries between the Earth's major convective air masses are themselves unstable, embroidering the atmosphere with large-scale wind eddies that swirl outward from localized high pressure areas, and inward toward areas of surface low pressure (Figure 4.2). On a flat Earth the surface winds would be expected to flow directly from areas of high pressure to areas of low pressure, in order to reduce the pressure gradient between the two locations. The Coriolis effect acting on the surface of a rotating sphere greatly complicates matters, however, causing an apparent deflection of wind currents to the right (in the Northern Hemisphere) or left (in the Southern Hemisphere). Rather than taking the direct route, air molecules must crab-walk sideways toward their destination. Northern Hemisphere winds thus spiral clockwise away from areas

Figure 4.2. Snapshot of actual wind flow patterns in the Pacific Ocean.
Note convergence of trade winds slightly north of the equator, and tight
spirals of low-pressure systems in the southern Pacific (based on satellite radar
measurement of wave orientations by NASA).

of descending, high-pressure air and counterclockwise into areas of
rising, low-pressure air.

    Surface winds therefore reflect both the large-scale atmospheric
circulation motivated by planetary-scale temperature differences and
regional patterns of pressure-driven flow. The former can establish reasonably reliable wind resources, such as the gentle westerly winds that
prevail in middle latitudes and the consistent northeast trade winds
felt most of the time by islands in the Caribbean. The latter are by
definition unstable, resulting in local winds that continuously change
in direction and intensity as pressure centers and associated weather

fronts march across the Earth's surface. These winds are often strong but are less consistent, and at times can become downright violent.

The strongest planetary winds blow at high altitude, near the top of the troposphere at elevations in the range of 10 km. Called jet streams, these meandering, fast-moving rivers of air can reach velocities of 300 km per hour or more. Jet streams are found at the boundaries between the major atmospheric circulation cells (Figure 3.5), where high-altitude airflow is deflected eastward by Coriolis forces. Their extreme velocity results from the air temperature contrast between the relatively warm air nearer the equator and relatively cold air nearer the poles, combined with the acceleration that inevitably occurs when air flows toward the poles. The latter effect is directly analogous to a spinning figure skater, who spins slowly when her arms and legs are extended to sweep out large circles, but who speeds up when they are pulled in close. Because the Earth is round, air moving toward the poles simultaneously moves closer to its axis of rotation, which is the equivalent of a skater's tightening her spin.

Can humans tap into wind energy flowing miles overhead? After all, today's largest wind turbines reach upward only about 2 percent of that distance. The tallest building in the world, currently the 828 m Burj Khalifa in Dubai, only reaches about 10 percent. On the other hand, oil wells have already drilled down into the Earth a distance equal to the elevation of the jet streams above. High-altitude wind power may not be quite so outlandish as it first sounds. International conferences have already been held on the subject, attended by serious scientists and engineers with impressive credentials and ideas for high-altitude kites and airborne wind turbines. Who knows what might result? In fact we already make some use of the energy in jet streams, through judicious planning of jet aircraft routes. Intercontinental flights eastbound can cut an hour or more off the trip if they can hitch a ride in a jet stream, with proportionate savings of fuel. For most of human history this use of the wind would have been literally unimaginable; the discovery of the jet stream and the invention of the jet engine both occurred in the mid-20th century. Both innovations now seem drearily routine to airline passengers, who tend to rank legroom, movie selection, and the line for the lavatory higher in their list of concerns.

The Edge of the Air: Wind Near the Earth's Surface

If the Earth's surface were perfectly smooth and frictionless, wind turbines could be economically built anywhere that global or regional wind patterns allow. However, the surface is actually quite rough, greatly complicating human use of wind energy. Mountains, hills, trees, waves, buildings, billboards, cars, cows, and other surface irregularities all conspire to obstruct low-level winds, diverting them from their appointed rounds. Local areas of rising warm air called thermals do the same, by adding a vertical component of motion. Collectively these disruptions contribute to turbulence, which shows itself as localized wind currents that differ in speed or direction from the main body of wind. The turbulent mixing of air is most intense close to the ground and typically decreases upward.

Early in the history of aviation some unknown aeronautical expert recommended that pilots try to "stay in the middle of the air," and "don't go near the edges of it." Pilots of small planes flying near the lower edge of the air (close to the ground) know that the air is often bumpy there, but that if they climb higher the bumps become less severe and often disappear entirely. The lower bumpy part of the atmosphere has been named the planetary boundary layer, because it is directly in contact with the planet (Figure 4.3). Above the planetary boundary layer the rest of the atmosphere goes about its business relatively unimpeded by the friction generated below. The top of the planetary boundary layer is often marked by clouds, making it readily visible. Experienced pilots wait until they are above the clouds to pour their first cup of hot coffee. The thickness of the planetary boundary layer varies considerably, from less than 100 m up to 3,000 m. It grows thicker over rough land surfaces with lots of thermals and thins over the sea.

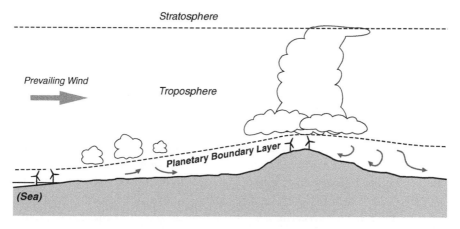

Figure 4.3. Schematic illustration of the planetary boundary layer.

The net effect of boundary layer turbulence is to slow down winds near the Earth's surface, since turbulent eddies are working at cross-purposes to the overall flow. Unfortunately wind turbines must operate within the planetary boundary layer, which limits their potential. The worst effects occur nearest the surface, with effective wind speed increasing with height. This basic fact has led to the construction of ever-more-colossal wind turbines that can reach higher into the planetary boundary layer. The world's largest wind turbines reach upward nearly 200 m at the top of their arc and are visible up to 50 km away. Taller wind turbines present increased engineering challenges, however, in part because of their sheer weight and required support structure. Gigantic turbine blades are also susceptible to imbalance, since they experience stronger winds at the top of their arc than at the bottom, which can cause premature wear and tear. A variety of other practical limitations also come into play; for example, very tall turbines become hazards to aerial navigation. They must therefore be lighted to warn off aircraft, further boosting their visual impact.

The Topography of Wind

An alternative (and complementary) approach to the brute-force solution of building ever-taller turbines is to build them in places where the planetary boundary layer is thinner. A thinner boundary layer means that stronger winds blow closer to the ground. The vertical wind gradient within the boundary layer therefore becomes more pronounced, and wind speed increases more rapidly with height. A tower constructed where the boundary layer is thin will therefore produce more power than the same tower constructed where the boundary layer is thick (other factors being equal).

Generally speaking the boundary layer is thinner where surface obstructions such as trees and topography are minimal. Flat plains often experience strong winds nearer the land surface and therefore can be prime areas for wind development. The central plains of North America in particular are renowned for their windiness. The plains are underlain by an assemblage of continental rocks greater than 1 billion years old, collectively known as a "craton." Nothing very geologically exciting has happened there for hundreds of millions of years, allowing the topographic highs of the craton to be gradually worn down by erosion. Deposition of flat-lying sedimentary rocks has simultaneously filled in the lows. The rest of the central plains of the United States have experienced similarly prolonged boredom and are often referred

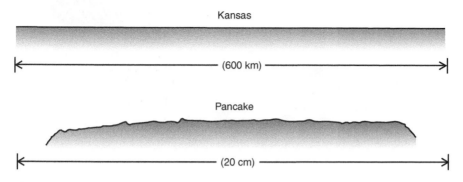

Figure 4.4. Comparison of Kansas topography to that of a pancake (modified from Fonstad et al., 2003).

to as being "flat as a pancake." A study of Kansas published by Mark Fonstad, William Pugatch, and Brandon Vogt in the somewhat less than prestigious journal *Annals of Improbable Research* demonstrated that the topography of Kansas is in fact considerably *flatter* than a pancake, proportionately speaking (Figure 4.4). Flattened, windy cratons form the core of several other continents as well, most notably interior areas of South America, Australia, and parts of Africa.

If wind speeds increase with elevation, why not site wind farms in mountainous areas rather than on low-lying plains? Mountaintop winds can attain truly ferocious intensity; for example, the strongest officially recorded wind was a gust of 372 km/hr at the summit of Mount Washington in New Hampshire. To some extent mountainous regions do offer useful wind resources, particularly in locations such as mountain passes that force winds to blow through a narrow gap. Like water flowing through the constricted nozzle of a garden hose, the wind must speed up through a mountain pass to maintain its overall flow rate. Much the same venturi effect can be felt above smoothly sculpted mountain ridges, where the planetary boundary layer tends to squeeze itself thinner to flow over the underlying topography without disturbing the overlying air mass.

A variety of factors conspire against extensive development of wind resources in rugged mountainous regions, however. For starters, air density decreases with altitude; at 3,000 m elevation air density is only about 70 percent that of sea level. Wind blowing at that elevation therefore possesses only 70 percent as much kinetic energy as wind blowing at the same speed over low plains. Rugged terrain also tends to induce strong and chaotic turbulence. Consistent winds

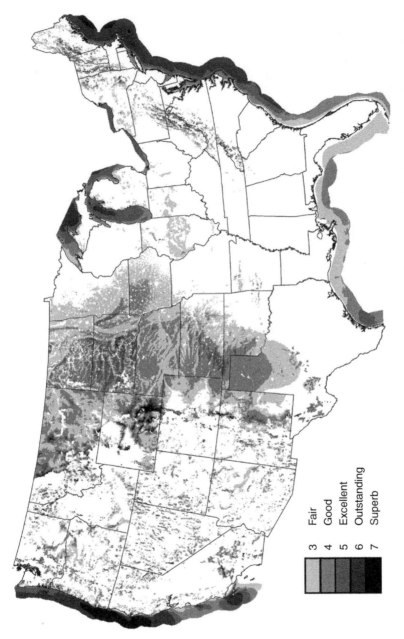

Figure 4.5. Wind resource map of the United States. Note that highest winds correspond to the flat topography of the Great Plains (U.S. Dept. of Energy, National Renewable Energy Laboratory).

3	Fair
4	Good
5	Excellent
6	Outstanding
7	Superb

of moderate strength are therefore elusive. There are also practical difficulties associated with installing and maintaining large turbines that can withstand gale-force peak winds on terrain that might make mountain goats nervous. Adjacent mountain valleys pose fewer problems, but are more sheltered and commonly experience temperature inversions that inhibit air movement.

The best quality wind energy resources on the continents are found on the highest parts of the flattest terrain, where relatively smooth, low-lying ridges extend upward into an already thin boundary layer. A variety of geologic features can form such ridges. For example, the high plains of the central United States are underlain by an eastward-sloping wedge of sediment derived from volcanoes in the Rocky Mountains between about 40 and 25 million years ago. Later river erosion cut valleys into these deposits, leaving behind flat-topped mesas that experience particularly windy conditions (Figure 4.5). Subtle bedrock ridges such as the Niagaran escarpment in the Great Lakes region provide similar topographic advantages, as do the gentle landforms left behind by glaciers. Such features may rise only tens to hundreds of feet above the surrounding landscape, but this is enough to improve the quality of wind resource substantially.

Sea Breezes: Offshore Wind Energy

The most obvious place to look for a thin planetary boundary layer is at sea, where surface roughness is measured in meters of wave height, and thermals are virtually nonexistent. The temperature contrasts between land and sea also drive coastal breezes that help ensure a generous supply of wind. Maps of available wind resources invariably rank coastal waters as having higher potential than nearly anywhere onshore. So why not build wind turbines there? At this writing several European countries have already done just that, and it seems only a matter of time before other regions follow suit. Building wind farms offshore does inevitably require a larger infrastructure investment, but this investment may be justified by the greater energy returns.

How far offshore can we go? Currently most wind development is restricted to areas fairly close to shore, in water depths of 30 m or less. Staying close to shore minimizes development cost, good news for entrepreneurs looking to turn a profit. However, it also puts turbines in the direct line of sight of beachgoers and other coastal residents long accustomed to gazing out on an open sea. The Cape Wind project in Massachusetts thrust this conflict into the limelight in the early 2000s, when objections were raised to a planned project for placing turbines 160 m tall in environmentally sensitive Nantucket Sound. A shrill debate soon erupted over both the environmental impact and the true motives of opposition, who included a number of wealthy individuals with beachfront property.

The coastal configuration of North America and other continents suggests a compromise solution that might prevent such controversies: Go farther offshore. The continents are rimmed by relatively shallow seas called continental shelves, which extend tens to hundreds of kilometers seaward. During the peak of the last ice age these areas were in fact dry land. Subsequent melting of continental-scale glacial ice sheets released enormous amounts of water, flooding the outer continental shelves to depths exceeding 100 m. Oil production platforms colonized the continental shelves of the Gulf of Mexico, the North Sea, and many other areas decades ago; there is no a priori reason why wind turbines could not follow in the future. The energy potential of offshore wind is impressive; for example, a 2006 report by the Minerals Management Service estimated that 900,000 MW could be generated on the continental shelves adjacent to the United States; that quantity represents roughly twice the actual usage of electricity in the United States in 2009. The Netherlands has targeted an installed capacity of 6,000 MW in the Dutch sector of the North Sea by 2020, which equals about half of that country's' current electricity usage.

Past the edge of the continental shelves, ocean depth increases more rapidly, to a global average of approximately 4,000 m (the maximum depth reaches nearly 11,000 m). Development of wind resources in the open ocean may still be technically feasible; oil exploration and production have already reached approximately 2,400 m water depth. However, the costs of building and operating wind farms in such deep areas would likely increase exponentially.

River Power

Roughly one-quarter of the solar energy reaching the Earth's surface is employed in converting water from a liquid to a vapor, through the process of evaporation. Solar heating imparts kinetic energy to water molecules at the surface of the ocean (or other water bodies), allowing them to enter a gaseous state and mix with the overlying air. The water vapor thus mixed is invisible to the eye and is sensible only as increased humidity. Evaporation has its limits, of course; otherwise the ocean would have dried up long ago and fish would all be homeless. The reason they are not is that the capacity of air to hold water vapor is strictly regulated by temperature: The higher the air temperature, the more water vapor can be held. The atmosphere is presently far too cold to hold an entire ocean of water.

About 90 percent of the water vapor that does reach the atmosphere comes from evaporation of the ocean; the remainder results from evaporation from water bodies on land and from plants (called transpiration). If all of this water vapor could be instantly forced back to liquid form, two things would happen. First, most of the planet would be drenched in a brief but torrential rainfall (or snowfall), averaging about 2.5 cm in total. Good for the garden perhaps, but not enough to cause a Biblical flood. Second, the kinetic energy that was originally imparted to water molecules by the Sun would be converted back to heat. In a sense this heat never really went away. Like latent fingerprints at a crime scene, it was there all along but not readily detectable. Latent heat (its actual name) is routinely released when water vapor condenses into clouds, which are formed from small water droplets that reflect light. The total latent heat of the atmosphere is enormous; a rough calculation shows it to be about equal to the energy content of all currently known fossil fuel reserves. Even more surprising, this latent heat is renewed roughly every ten days, the average amount of time that elapses between evaporation of a water molecule and its eventual incorporation into a water droplet.

The release of latent heat during condensation helps to propel moist air upward in the atmosphere, by adding to the original solar heating of air. At the extreme this process helps drive thunderstorms in which clouds can reach heights as great as 20 km. As it rises, water in the atmosphere trades thermal energy for gravitational energy, at a rate that can be expressed by a simple formula:

$$U = mgh$$

U in this equation is gravitational energy, a form of potential energy possessed by any object (or atom) not located at the center of the Earth; m is the mass of that object (expressed in kilograms), and h is its distance from the center of the Earth (in meters); g is simply a constant (9.8 meters per second, squared) that measures the strength of the Earth's gravitational field.

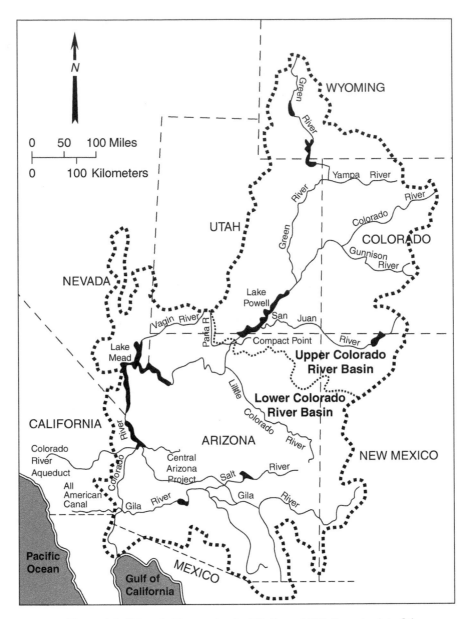

Figure 4.6. Colorado River watershed (R. Ferrari, U.S. Department of the Interior, Bureau of Reclamation, 2001).

Note that precipitation falling across a large geographic area is focused onto two dams, the Glen Canyon Dam (Lake Powell) and the Hoover Dam (Lake Meade) (Modified from Ferrari, 2008).

This gravitational energy can be traded back into kinetic energy if the vapor condenses into droplets and falls from the sky as precipitation. Raindrops splashing into the sea convert their kinetic energy back to heat, closing the circle that began with solar heating. However, precipitation that falls on land remains temporarily trapped at some elevation above sea level. It must hitch a ride with a river to complete its journey to the sea. Rivers therefore act as a sort of continental-scale circulatory system, or alternatively as machines whose job is to gather up water, sediment, and dissolved chemicals from across the landscape and deliver them to the ocean.

Rivers also focus the diffuse gravitational energy of continental precipitation into narrow streams that can be conveniently exploited by humans (Figure 4.6). The potential magnitude of this resource can be estimated by multiplying the mass of water that flows through the world's rivers by the gravitational constant and by the average elevation of the land surface. This calculation suggests that the global power of rivers nearly matches our current overall rates of global energy consumption. River power looks even more promising if you consider that precipitation is not evenly distributed across the land surface, but instead tends to fall in disproportionate quantity at higher elevations.

In actuality rivers supply only about 2 percent of the world's total energy consumption, and this percentage has remained relatively constant in recent decades. While the total magnitude of the resource may be large, certain practical realities intrude on its utilization.

Double-Edged Dams

Hydroelectric power works by converting some of the gravitational potential of precipitation into kinetic energy, which can in turn be used to drive an electric generator. For river water to give up its gravitational energy it must give up elevation, something it does naturally and continuously as it flows toward the sea. The natural slope of riverbeds is usually quite gentle, however; for example, the upper Yangtze River in China loses only about 25 cm per km on average. Rivers must therefore travel a surprisingly long horizontal distance to achieve much elevation drop, making it awkward to capture much useful energy.

This obstacle is not insurmountable, however, and humans have successfully used river power for thousands of years. One way to do so is to construct a smooth channel that captures part (or all) of the upstream flow of a river, carries it downstream at a lower gradient than the river itself, and then discharges it through a waterwheel or turbine. This approach can be effective, but it requires large investments in infrastructure. Generally it is only practical in steep, narrow mountain valleys and for relatively small amounts of electricity generation. The power output of such systems depends directly on river discharge, which in mountainous regions can vary by orders of magnitude in response to local, short-term variations in precipitation.

Hydroelectric dams, on the other hand, require only that an obstruction be built at one point of the river, to a height equal to the desired elevation drop. The reservoir created by a dam also helps to dampen water supply variations, because the total volume of stored water is large compared to the discharge rate. Single storms or short-lived dry spells are of little consequence, although long-term drought may eventually cause reductions in generation capacity.

Dams are perhaps unique among energy systems in that they can provide a host of other benefits that are not directly related to power generation. In many cases their primary value is to ensure a consistent water supply for irrigation or residential use in semiarid regions or other areas with strongly seasonal precipitation. For example, the Nile River in Egypt naturally tends to flood in late summer to early fall, in response to seasonal rainfall on the Ethiopian highlands. Completion of the Aswan High Dam in 1970 alleviated this flooding and provided a stable water supply for irrigation during droughts. Dams in the western United States capture spring snowmelt from high mountains, so that it may be used in later seasons. The reservoirs associated with dams are also a recreational boon to boaters, water-skiers, fishermen, and other lacustrine enthusiasts. Artificial lakes are especially welcomed by real estate developers, because flooding of preexisting river valleys creates far more waterfront property than would exist on a natural lake of similar area.

Dams are also subject to certain practical limitations, however. They really only work in areas where cooperative topography has already done most of the job of constricting a river, leaving only a small gap to be closed by humans. The Hoover Dam, for example, was built across a narrow canyon cut by the Colorado River between what are now the states of Nevada and Arizona. It only needs to span a gap of 379 m to raise its reservoir 221 m above the level of the downstream Colorado River. Lacking such natural topographic assistance, dams would have to be prohibitively lengthy. For example, the lower Mississippi River flows across generally flat terrain on its way to the

Gulf of Mexico. A dam built across the Mississippi to the same height as the Hoover Dam would need to stretch from Little Rock, Arkansas, to Nashville, Tennessee, a distance of more than 500 km. The resultant lake would flood most of the central United States. Because many of the world's larger rivers flow across relatively flat terrain, much of their energy is effectively off-limits to exploitation.

Far from providing free energy, hydroelectric dams and their associated reservoirs impose environmental costs that are both substantial and complex. The most immediate impact is inundation of land surface area by reservoirs, which can carry significant human cost. Lake Nasser in Egypt is estimated to have forced the relocation of at least 100,000 people, and more than 1 million people had to be relocated during construction of the Three Gorges Dam in China. Historical sites and cultural landmarks in the path of the floodwaters generally cannot be moved and are therefore flooded, a gift perhaps to future archaeologists.

Ironically, dams built to ensure stable water supplies in dry regions inevitably squander part of that supply, through evaporation from the surface of the newly created reservoir. For example, the volume of water evaporating each year from Lake Meade, the reservoir impounded by the Hoover Dam, is comparable to the total annual water usage of the City of Los Angeles. Lake Nasser, lying in the Saharan desert, loses as much as 20 percent of its inflow from the Nile River to evaporation. The lake has been effective at controlling floods and has permitted an expansion of agriculture. It has also damaged soils, however, through buildup of salt. The natural flooding cycle caused groundwater levels to fluctuate, a process that tends to limit salt accumulation because the soils are flushed with freshwater during floods. However, with downstream Nile River flow reduced by the dam, groundwater levels have remained consistently low, allowing salts to build up in the soil through continual evaporation of irrigation water. Dams also drastically alter the ecology of river systems, both upstream and downstream, in myriad ways that are difficult to predict and remediate.

All artificial reservoirs are ephemeral, fated to eventual defeat at the hands of the rivers they obstruct. Slow but relentless infill by sediment brings about this destruction, eventually filling reservoirs completely. This occurs because rivers slow when they enter a standing reservoir, losing their capacity to carry sediment. Silt and clay are initially dumped into a delta near the upstream end of the reservoir (Figures 4.6 and 4.7), and over time this delta expands into muddy flats that ultimately reach the dam. The global magnitude of this infill is impressive; geologist

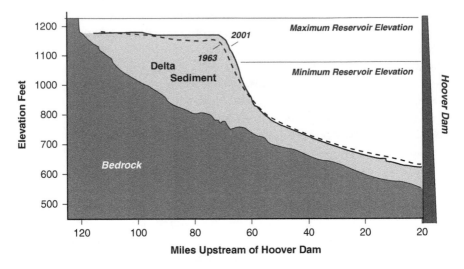

Figure 4.7. Longitudinal profile of Lake Meade along the path of the Colorado River, upstream of the Hoover Dam.

Note that perhaps 40 percent of the original river valley filled with sediment between construction of the Hoover Dam in 1935 and 1963, and about 7 percent of the original reservoir volume was lost. Construction of the upstream Glen Canyon Dam in 1963 dramatically slowed sedimentation in Lake Meade since (sediments are now trapped in Lake Powell) (modified from Ferrari, 2008).

James Syvitiski and his colleagues in 2005 estimated that 20 percent of the global riverine sediment load is currently being captured by large reservoirs, and another 23 percent may be trapped by smaller impoundments. The time required for individual reservoirs to fill up varies with size of the reservoir and sediment discharge from the river; the global average appears to be on the order of five hundred years.

With their reservoirs filled by sediment, dams effectively become concrete waterfalls. Continued maintenance of these waterfalls will still be required even after they lose their usefulness, to ensure that their trapped sediment remains trapped. The impounded sediments might otherwise become a sort of environmental abscess, awaiting its chance to burst and spill the accumulated sediment, agricultural fertilizers, pesticides, and mining and industrial wastes of several centuries. Most of the world's major dams are relatively young, built during the 20th century, and no clear exit strategy has been devised for dealing with their old age.

The trapping of sediment by dams can also result in negative consequences for downstream topography. Lake Nasser, for example, traps sediment that formerly was transported to the Nile Delta, a giant pile of mud and sand thousands of meters thick. The weight of sediment at the top of the pile continually compresses sediment below, resulting in slow, relentless subsidence of the delta surface. The fertile delta would disappear entirely beneath the eastern Mediterranean Sea were it not continually replenished by newly transported river sediment. Sediment trapped behind the Aswan Dam cannot replenish the delta, however, which has already begun to lose its battle with the waves.

The Motion of the Ocean: Energy from Waves and Tides

Waves offer yet another way to recover energy from moving water. The largest waves ever recorded have a height of around 20 m, but typical fair-weather waves in the open ocean are closer to 1 m. Assuming a typical wavelength of about 100 m, wave crests are regenerated about every twenty seconds. If you could somehow stack each wave crest one on top of the other for a period of twenty-four hours, you would have a pile of water approximately 4,300 m tall, equal to twenty times the height of the Hoover Dam. The total power of waves impacting a coastline therefore is immense, but the efficient collection of energy that arrives in such small parcels presents significant engineering challenges.

Waves are a bit more complex than this simple analysis suggests, however, because of the way water moves as a wave passes. Rather than simply going up and down, it simultaneously moves back and forth, so that particles in the water ultimately travel in a complete circle (Figure 4.8). For practical purposes wave power can be summarized by a relatively simple equation:

$$Power = T \, H^2 \times (constant)$$

The T in this equation is wave period, equal to the number of seconds between successive wave crests; H is wave height, equal to the vertical difference between crest and trough. Note that H is squared, making it the dominant control on wave power. These quantities are then multiplied by a constant, which takes into account factors such as water density and the strength of the gravitational field.

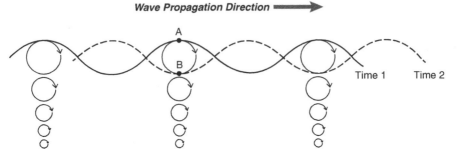

Figure 4.8. Schematic illustration of water movement associated with waves traveling along a water surface.

Circles indicate the paths taken by floating objects as each wave passes; note that wave influence decreases with depth. A = height of wave crest, B = height of wave trough.

Wind-driven waves arise because wind and water both have viscosity (a measure of fluid "stickiness"), which allows energy to be transferred by shear across their boundary. Small instabilities on this surface become more organized waves, which grow in proportion to the speed of the wind and the distance (fetch) over which it blows. Waves therefore tend to be larger on the downwind side of large water bodies, other factors being equal. The largest waves are generated by major storms, but once generated they can travel far beyond their point of origin. Sun-tanned surfers in Hawaii, for example, enjoy a coincidence of sunny weather and large waves, which originated from storms half an ocean away, near New Zealand.

Wave height is further influenced along irregular coastlines, by the bending of wave fronts as they approach the coast. This happens because in shallow water, wave propagation speed is slowed by interaction with the seafloor. The water within a single wave therefore slows when it begins to interact with the shallow seafloor, but maintains its original speed over deeper areas. The net result is that wave fronts turn to become more parallel to shore (Figure 4.9). This in turn causes waves to focus their energy on headlands. The opposite happens in coastal bays, where gentler waves allow the accumulation of sandy beaches.

Tides occur because of gravitational attractions that distort the shape of the ocean's surface. Although we tend to think of the Moon as orbiting the Earth, in reality both bodies orbit their common center of mass, located just below the Earth's surface. Two major sea-surface

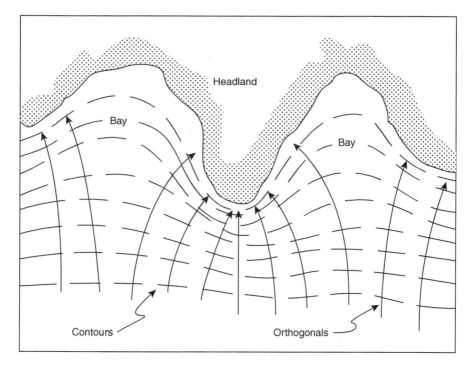

Figure 4.9. Schematic illustration of wave focusing on headlands (U.S. Army Corps of Engineers).

bulges result, one pointed toward the Moon due to direct gravitational attraction, and an opposite bulge caused by the tendency of the ocean to be flung away from the orbital center of mass. The Sun exerts a somewhat weaker tidal force, because it is located much farther away. The Sun's tidal attraction reinforces that of the Moon when all three bodies fall in a line (spring tide) and competes with the Moon when the Sun and Moon are at 90° relative to the Earth (neap tide). The tide rises and falls at a given location as the Earth rotates relative to the gravitational distortions of the ocean surface.

Because they have much longer wavelengths than wind-driven waves, tides enjoy additional natural amplification by certain coastal geometries. Specifically, tidal range tends to increase along concave-shaped coastlines and in V-shaped embayments. The reason is fairly simple: As incoming tidal currents are squeezed between the sides by the embayment, the water has nowhere to go but up. The most famous example of this is probably the Bay of Fundy in Nova Scotia, which becomes progressively narrower inland (Figure 4.10). At

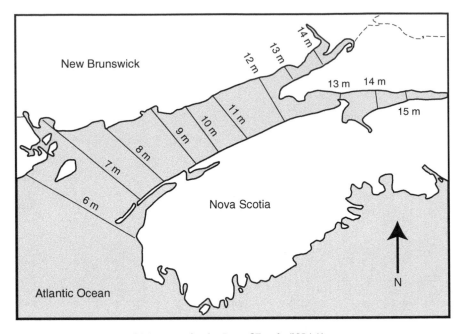

Figure 4.10. Tidal ranges in the Bay of Fundy (NOAA).

its inland extreme, tidal range reaches as high as 15 m, which represents a dramatic increase over open ocean tides of less than 1 meter. The Bay of Fundy is floored by immense mud and sand flats that are violently flooded twice per day, potential deathtraps for the careless or the unaware.

An impressive diversity of designs have been proposed for machines that collect the energy of waves and tides, some of which go back centuries in origin. Practical wave power remains stuck in its infancy, however. The first modern commercial wave farms, built in Portugal in 2008, quickly succumbed financially in a poor world economy. Additional projects are in the development or test phases in a number of other countries, but it remains to be seen whether these will ultimately prove practical.

Conclusions

It is often said that the three most important factors in real estate are location, location, and location. The same can be said of wind, rivers, waves, and tides as energy sources. Collectively they contain an enormous quantity of renewable energy, but most of this energy is spread too thinly across the enormous surface area of the Earth to be useful. For these energy sources to be practically exploited they must be naturally concentrated. Generally speaking this means that the availability of renewable energy from the fluid Earth depends directly on the geologic evolution of the solid Earth, which has shaped the world's coastlines, deserts, windy interiors, and mountainous river valleys.

Because of this requirement for natural geographic concentration, energy from the fluid Earth is not always found where we want it. For example, wind energy is abundant in North Dakota, where people are relatively scarce. In such cases the effective cost of renewable energy increases, because of the substantial infrastructure required to transport it where it is needed. In other cases the geography is more fortuitous; for example, much of the world's population lives near coastlines where wind, wave, or tidal power may be amply available.

For More Information

Ahrens, C. D., 2007, *Meteorology Today: An Introduction to Weather, Climate, and the Environment*, 8th ed.: Belmont, CA, Thompon Higher Education, 537 p.

Archer, C. L., and Jacobson, M. Z., 2005, Evaluation of global wind power: *Journal of Geophysical Research*, v. 110, D12110, DOI:10.1029/2004JD005462.

Boyle, G. ed., 2004, *Renewable Energy: Power for a Sustainable Future*: Oxford, Oxford University Press in association with the Open University, 452 p.

Ferrari, R. L., 2001, Lake Meade Sedimentation Survey: U.S. Department of the Interior, Bureau of Reclamation Report, 212 p.

Fonstad, M., Pugatch, W., and Vogt, B., 2003, Kansas is flatter than a pancake: *Annals of Improbable Research*, May/June, p. 16–18.

Isaacs, J. D., and Schmitt, W. R., 1980, Ocean energy: Forms and prospects: *Science*, v. 207, p. 265–273.

Manwell, J. R., McGowan, J. G., and Rogers, A. L., 2002, *Wind Energy Explained: Theory, Design, and Application*: Chichester, UK, John Wiley and Sons, 577 p.

McCully, P., 2001, *Silenced Rivers: The Ecology and Politics of Large Dams*: London, Zed Books, 359 p.

Syvitski, J. P. M., Vörösmarty, C. J., Kettner, A. J., and Green, P., 2005, Impact of humans on the flux of terrestrial sediment to the global coastal ocean: *Science*, v. 308, p. 376–380, DOI: 10.1126/science.1109454

Vörösmarty, C. J., Meybeck, M., Fekete, B., Sharma, K., Green, P., and Syvitski, J. P. M., 2003, Anthropogenic sediment retention: Major global impact from registered river impoundments: *Global and Planetary Change*, v. 39, p. 169–190.

5

Covered in Green: Biofuels Basics

> But of all the occupations by which gain is secured, none is better than agriculture, none more profitable, none more delightful, none more becoming to a freeman.
>
> *Marcus Tullius Cicero, De Officiis (44 BCE)*

Biofuels built Rome! The motive force behind Rome and every other preindustrial civilization came almost entirely from muscle (human and animal) and was thus fueled by agricultural food production. Cooking and heating were mostly accomplished by directly burning wood and other plant matter. During the Industrial Revolution we gradually moved away from biological energy sources, initially for practical reasons related to the greater energy density offered by coal and petroleum. Coal, for example, provides about twice the energy contained in an equal weight of biomass, and about eight times as much energy by volume (Figure 5.1). Oil and gas offer even larger advantages. Fossil fuels also reduced the need for human and animal labor, since higher quality fuels make mechanization more practical.

Despite its historical decline in popularity, energy derived from recently living plants continues to offer two potentially key advantages over fossil fuels. First, living plants continuously recycle carbon dioxide from the atmosphere, a process that in principle means their use should cause no net change in greenhouse gas abundance. Second, plant biomass is renewed with each new growth cycle, implying that biofuels should always be available (with certain important caveats). Increasing concern over both the future availability of fossil fuels and their inevitable net addition of CO_2 to the atmosphere has therefore led to renewed interest in biofuels.

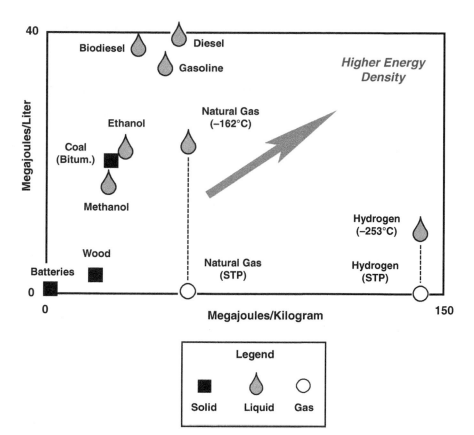

Figure 5.1. Energy density.
The horizontal axis gives energy per unit mass; the vertical axis gives energy per unit volume (STP = standard temperature and pressure at sea level).

The most immediate questions for biofuels are whether their potential advantages can be achieved without sacrificing the well-known advantages of fossil fuels, and whether industrial-scale biofuel production can be accomplished without threatening food supplies or the environment. These questions are by nature complex, and informed debate requires a deeper grasp of the inner workings of biological energy systems and their interactions with the Earth. This chapter therefore provides an introduction to biofuel basics, starting with the evolution of plants and ending with their conversion into high-quality fuels.

A third, equally important question is whether the production of renewable biofuels will consume natural resources that may not be renewable (soil, nutrients, and freshwater). This question will be taken up in the next chapter.

Catching Rays: The Evolution of Photosynthesis

Green slime rarely gets the respect it deserves. To the casual observer algae appear to be the lazy vagabonds of the plant world, floating list-lessly across the surface of ponds, lakes, and shallow seas. They seem to have no real ambition in life beyond basking in the Sun, or per-haps providing a haven for fish. Most algae are not even able to stand upright unassisted, as do their more promising cousins the grasses, shrubs, and trees. However, this relaxed façade conceals a beehive of inner exertion, including the rather miraculous task of transforming common household materials into the basic foodstuff of life:

$$\text{Carbon dioxide} + \text{water} (+ \text{light}) \rightarrow \text{sugar} + \text{oxygen}$$

Sunlight powers this transformation and along the way is partly con-verted into chemical energy that may be stored for later retrieval. This process, called photosynthesis, also releases oxygen as a by-product, allowing us to breathe. Not a bad day's work all told!

Primitive organisms called cyanobacteria learned this trick very early in the history of the Earth, 3 billion years ago or possibly even earlier. The Earth at the time was a hellish place compared to today, with an atmosphere dominated by volcanic emission of water vapor and poisonous gases such as carbon dioxide, ammonia, and hydrogen sulfide. Between 2.5 and 2 billion years ago cyanobacteria became so successful that the atmosphere contained appreciable amounts of free oxygen. The availability of atmospheric oxygen also allowed ammonia (NH_3) to be converted to gaseous nitrogen. Between about 1 billion and 500 million years ago cyanobacteria franchised their technology to the more biologically advanced algae, and business has continued to improve ever since. Algae have progressively diversified, with many new species that are grouped into green, brown, and even red varie-ties. In addition to producing an estimated 70–80 percent of the oxy-gen we breathe, algae provide the foundation for the entire marine food chain.

The seemingly more ambitious land plants actually arrived rather late to the job, not appearing in abundance on the continents until around 450 million years ago (only about the last 10 percent of Earth history). Fungi may have already been there to greet them but failed to register an indisputable claim in the fossil record. Why did it take so long to colonize the continents? This question is not easily answered, since most of the evidence is long vanished. It was presumably a quiet invasion, starting at the coasts or other low-lying areas near open water. The earliest plants were seedless and therefore needed waterlogged lowlands to spread. This restriction was lifted with the later evolution of seeds, which allowed plants to propagate inland across drier areas.

The first really explosive growth of forests occurred between about 350 and 300 million years ago and was partly responsible for creating the coal beds that eventually fueled the Industrial Revolution in Germany, Britain, and the United States. Those forests literally buzzed with animal life, ranging from the first reptiles to giant dragonflies with half-meter wingspans. Flowers were strangely absent, however, and remained so until around 140 million years ago. True grasses were an evolutionary offshoot of the flowering plants and appeared in abundance only about 40 million years ago. This family of plants, which has the Latin name Poaceae, counts among its members all of the major candidates for commercial ethanol biofuels, such as corn, sugarcane, and switchgrass. It also provides our primary food supply.

Life on the Rocks: The Geology of Soils

Part of the reason that plants took so long to colonize the continents may have been the difficulty of growing on barren rock. Rocks generally do not hold on to water very well and are not easily penetrated by root systems. They do contain nutrients needed by plants, but those nutrients are very tightly bound in crystalline form and thus not readily accessible. For plants, the early continents must have looked a bit like vacation property in the desert: cheap and abundant, but not much fun once you get there. Only committed enthusiasts tend to visit such places: for example, lichens that can grow on bare rock or plants that are able to extend roots into cracks and crevices and capture the small amount of available water. Neither group makes up a large proportion of the global biomass.

Fortunately for plants (and for humans), truly barren rock is a comparative rarity on the modern Earth. With relatively minor exceptions, exposed barren rocks are largely restricted to high mountains and deserts. Most of the land surface outside Antarctica and Greenland is covered with soils, which occur in dizzying variety. Numerous soil classification schemes have been devised, encompassing hundreds of distinct soil types. For example, the 1999 U.S. Geological Survey soil classification manual is 871 pages long and features such exotic soil names such as Fragiaquic Dystrudepts, Typic Torrifluvents, and Udic Kanhaplustalfs. To the uninitiated these names might as well be written in ancient runes. Ultimately they are based on different combinations of easily measurable characteristics, however, such as the development of soil layers, vertical soil structures, organic matter content, and soil chemistry. Soil classification can be represented in greatly simplified form by a mere dozen orders, which have been carefully mapped across the globe. Six of these together comprise two-thirds of the total soil surface area and virtually all potentially productive farmland (Figure 5.2).

At its essence, soil is the transition between rocks and the atmosphere, just as beaches are the transition between water and land. Soil forms through a wide variety of processes that are broadly known as weathering. The most dramatic of these processes resort to brute force to break big rocks into smaller rocks. Ice is one of the most effective natural agents of this destruction. Water seeps into open cracks in rock and then expands about 10 percent when it freezes. The resultant wedging pries the cracks farther open, and repeated through many freeze-thaw cycles ice can bring down mountains. Glaciers take matters to the next level, through active abrasion. Moving glaciers act as sandpaper, embedding rocks within the ice that help grind away at the surfaces over which they flow. The grindings, comprising enormous quantities of loose rock, sand, and fine silt, are dumped unceremoniously near the ice margin.

In regions without ice, other processes such as river erosion, landslides, and root wedging can similarly pummel rocks into submission but take longer to administer the beating. The net result is the same: to reduce originally intact bedrock to small broken bits of sediment. This sediment can then be picked up by rivers and be deposited on floodplains or carried farther downstream to be deposited where rivers meet the ocean (or lakes). Water delivers the coup de grâce by allowing the broken pieces to be fully or partially dissolved. The process is reasonably analogous to making coffee: The coffee beans are ground into smaller pieces, and then flushed with water. Grinding serves to increase the area of particle surfaces (mineral or bean) where dissolution may occur. Decaying plant matter in the upper layers of soil also produces organic acids, which further promote dissolution.

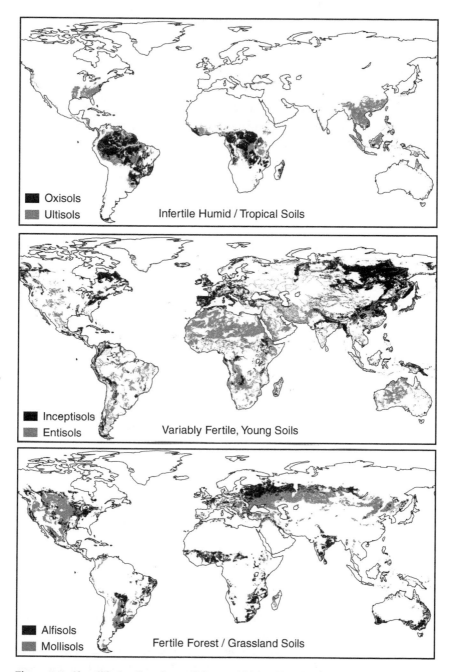

Figure 5.2. Simplified soil regions of the world (modified from U.S. Department of Agriculture, Natural Resources Conservation Service, Global Soil Regions map).

The elements released by dissolution of minerals may be precipitated as new minerals in deeper layers of the soil or be removed entirely by rivers and carried to the sea.

Soils soak up and hold water like sponges, providing a reliable drinking supply for plants. Plants return the favor by extending roots that help to hold the soil in place, protecting it from erosion. Plants also draw sustenance from the soil, in the form of a long shopping list of chemical elements needed to construct their tissues or maintain their metabolism. Nitrogen, phosphorus, and potassium lead this list and are considered primary nutrients. Nitrogen is used to build proteins and the molecules that carry a plant's genetic code. Nitrogen constitutes 78 percent of the Earth's atmosphere, but unfortunately atmospheric nitrogen is a self-loving gas. Each atom of nitrogen forms a triple chemical bond with another nitrogen atom, rendering it useless to plants. Some bacteria living in soil or symbiotically in plant roots possess the rare ability to convert atmospheric nitrogen to a form that plants can use and thus are the natural vendors of this important nutrient.

Phosphorus is used to build cell-wall membranes and combines with nitrogen to form the molecules that transfer the energy obtained from photosynthesis. Potassium serves as an electrolyte, enabling electrochemical reactions to occur in plants in much the same way as they do in our own cells. Humans need potassium for their muscle cells to function, and they get it by eating plants. Reading farther down the nutrient shopping list for plants we find calcium, sulfur, and magnesium ranked just behind the primary nutrients, and boron, chlorine, manganese, iron, zinc, copper, molybdenum, selenium as "micronutrients." Although not needed in large quantity, these lesser nutrients can nonetheless be showstoppers if they are lacking. Of this list of fourteen or so nutrients, all but nitrogen are derived from the weathering of rocks.

Global Patterns of Soil Fertility

Conventional wisdom suggests that plant growth depends principally on the availability of water, and that the potential for growing crops therefore should be highest where rainfall is the highest. This is partly true, and nowhere more so than in the lush rain forests found near the equator. In principle these areas would seem to have cornered the world market for biomass and by extension should ultimately control the availability of biofuels. However, the same warmth and rainfall that promote tropical rain forests also leave the underlying soils severely depleted in nutrients, limiting their value to agriculture.

Returning to the coffee analogy, imagine that rather than making one pot of coffee, you tried to make a dozen pots from the same grounds. Each pot would be successively weaker, because more and more essential coffee "nutrients" (most notably caffeine) are removed with each percolation. The same holds true for tropical soils: Increased flushing by rainwater and organic acids tends to dissolve and remove nutrients.

What is left is the stuff that does not readily dissolve, mostly clay minerals, quartz, and iron and aluminum oxides. Kaolinite is a particularly common clay mineral in tropical soils; its nutritional value to plants is nil. Quartz is a hard, shiny mineral that is likewise indigestible. Iron and aluminum oxides are also of little value to plants, although they do sometimes serve as ores for those metals. To add insult to injury, oxides also tend to sequester phosphorus and prevent its use by plants. Below the topmost organic-rich layer, humid tropical soils therefore consist mostly of thick layers of infertile red residuum. The overlying forest therefore depends utterly on nutrients that have accumulated in the standing biomass itself, and in dead organic matter lying below. "Slash-and-burn" agriculture exploits this delicate balancing act by collapsing the nutrients in the standing biomass back to the soil, temporarily permitting productive crop growth.

Soil orders associated with humid tropical conditions include oxisols and ultisols (see Figure 5.2). Oxisols have very little natural fertility, whereas ultisols are only slightly better. Sustained high-productivity crop growth is difficult to achieve without massive infusion of calcium carbonate (to reduce acidity and add calcium) and other nutrients, but with such additions these soils can be very productive.

At the other extreme, the floodplains of the Nile, Tigris, and Euphrates Rivers and their deltas have sustained rich agricultural production for millennia, despite being surrounded by desert. This is possible because floods continually replenish nutrients when they deposit new layers of sediment, and the rivers themselves provide a reliable source of water. The same is true of the historically productive floodplains of the Indus and Ganges Rivers in northern India and the Yellow and Yangtze River plains in eastern China. Together these seven rivers account for the majority of major ancient civilizations of the Old World. But what accounts for the rivers themselves? In a word, mountains. A single major domain of mountain ranges stretches from the Pyrenees in the west to the Himalayas in the east. This domain has no formal name, but all of its constituent ranges contain rocks deposited in an ancient equatorial ocean named Tethys (after a Greek goddess of the sea). The mountains were uplifted as this ocean closed, as a result of collision of continents and continental fragments that formed its southern margin with the Eurasian continent to its north. Two especially prominent continental collisions are ongoing today, involving India and Africa. The African collision uplifted mountains in eastern Turkey that are the source of the Tigris and Euphrates Rivers, as well as the Zagros mountains in Iran that feed tributaries of the Tigris. The Indian collision uplifted the Himalayas, which are the source of the Indus, Ganges, Yellow, and Yangtze Rivers (Figure 5.3).

Soil orders developed on recently deposited sediments or on recently exposed bare rock fall into the entisol and inceptisol orders. Because of their youth, these soils tend to be poorly developed and may lack distinct horizons. They are found in a wide range of settings, ranging from steep mountain slopes, to river bottoms, to deserts. The largest contiguous area of entisol is the Sahara desert, which consists mostly of windswept bare rock and sand. Needless to say, soil fertility is low. On the other hand, entisols and inceptisols found in river valleys can support highly productive agriculture.

The best soils in the world owe their existence to global change of the most radical sort: continental-scale glaciation and deglaciation. For the past 2.5 million years or so, gigantic ice sheets advanced dozens of times from centers in Canada, Greenland, Scandinavia, and Siberia. At their most ambitious moments, continental glaciers ventured as far south as Kansas and Missouri in the United States, across the entirety of the British Isles, much of northern Europe, and northern Russia (Figure 5.4). Today only Greenland remains glaciated in the Northern Hemisphere, a frigid reminder of the not so distant past. In their wake, continental glaciers left behind vast amounts of glacial sediment, that served as an ideal substrate for subsequent soil development. In some cases this material was simply dropped in areas that had been recently covered by ice, like equipment dropped by an army retreating in panic. The result is disorganized piles of sediment called *till*. In other cases streams of glacial meltwater sorted the sediment into more orderly deposits called *outwash*, which were commonly carried beyond the limits of the glaciers themselves. Finally, strong winds can pick up silt from both kinds of deposit and redistribute it as thick dust layers called *loess*.

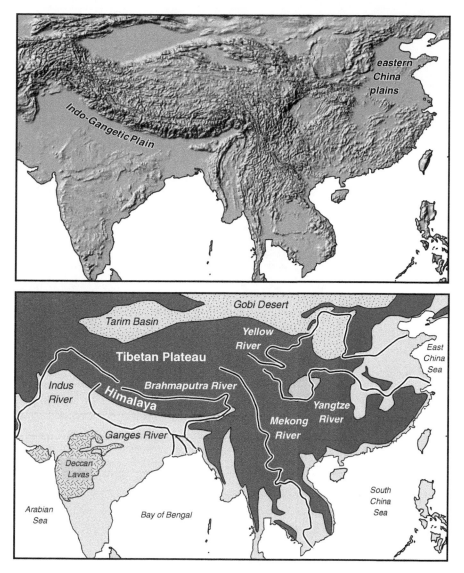

Figure 5.3. Top: shaded relief map of southeastern portions of Asia (Data source U.S. Geological Survey).

Bottom: simplified map of major rivers in the same region.

Dark gray = mountainous areas, stippled pattern = desert. Note that approximately 40 percent of the world's population lives within this map area.

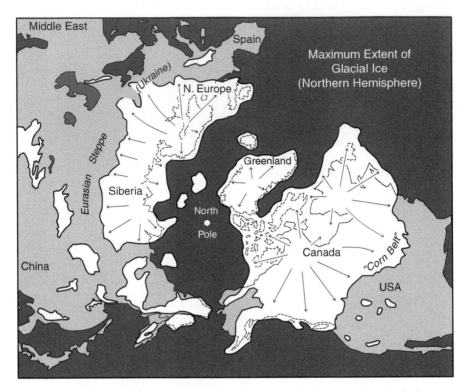

Figure 5.4. Maximum extent of Northern Hemisphere continental glaciers. Arrows depict the approximate direction of ice movement; ice generally flows away from areas of maximum accumulation and toward areas where it either melts or breaks up as icebergs. Isolated areas of glacier also occurred south of the major ice sheets, in areas of high elevation. The Corn Belt in the United States is mostly underlain by glacial sediments; the Eurasian Steppe is underlain by wind-blown dust (loess) derived in large part from glacial erosion (Modified from H. Grobe [2006, Wikipedia Creative Commons] and J. Schlee [2000, U.S. Geological Survey]).

The most productive agricultural regions enjoy a happy coincidence of several different factors, including adequate rainfall and climate; large, flat surface area; and good soil. To invoke the famous words of Goldilocks, everything is "just right." The Corn Belt of the central United States satisfies the Goldilocks principle as well as any place in the world (Figure 5.5). Rainfall there is moderate to moderately high. Flat land surface area is plentiful. For fifteen millennia the growth of prairie grasses enriched the glacial deposits, creating thick layers of fertile, organic-rich topsoil. The overall biomass potential of this area ranks among the highest in the world, and not surprisingly

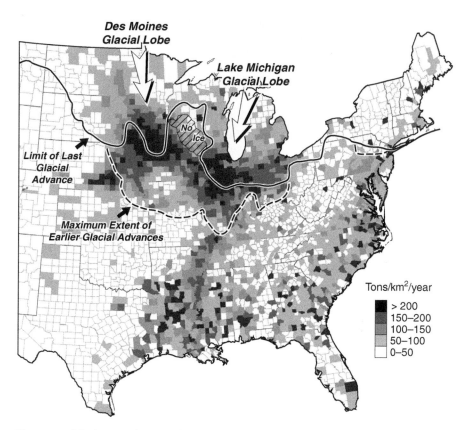

Figure 5.5. Maximum advance of glaciers in the eastern United States, and total biomass potential.

The Des Moines and Lake Michigan Lobes represent major areas of ice advance, as recorded by the position of glacial morraines. "No ice" represents an area between those lobes, which lacks glacial deposits. Biomass potential is the estimated total of agricultural, forest, and urban residues and wastes. Crop production is not included but follows similar large scale patterns. (Biomass data from Roberts, 2009, U.S. National Renewable Energy Laboratory).

it produces most of the biofuel presently used in the United States. Similarly enriched soils also exist in other locales that were visited by ice, including Britain, northern Europe, and Asia. Loess deposits extended the influence of the glaciers far beyond the limits of actual ice. The Eurasian Steppe consists of a vast tract of grasslands extending from Ukraine to Mongolia, which provided the principal corridor for the 13th century Mongol invasion of eastern Europe. It is underlain by rich soils, which are in large part formed on loess.

Alifisols and mollisols are the "just right" soil orders. They commonly form on glacial deposits and on other older sedimentary rocks and have developed naturally high fertililty. They are particularly common in areas of temperate climate, and most prevalent in the Northern Hemisphere.

It Isn't Easy Being Green: Plants as Solar Collectors

If you were to separate sunlight into a rainbow using a prism, you would find the brightest colors in the green part of the rainbow. Ironically, the characteristic green color of photosynthetic plants means they preferentially reflect green light rather than absorb it. Because of this, plants effectively waste the strongest part of the solar energy striking them. Black vegetation might seem funereal and depressing, but it would be considerably more effective at trapping sunlight across the entire color range. It would appear therefore that plants are not really designed to be efficient solar collectors; they gather only enough sunlight to meet the relatively modest energy requirements of their metabolism.

Plants can only look with envy at photovoltaic cells when it comes to collecting solar energy efficiently. Direct energy comparison of the efficiency of plants with photovoltaic cells is tricky, since plants do not directly produce electricity. One cannot, for example, plug a toaster directly into a cornstalk to measure how long it takes to make toast. However, the chemical energy that is stored in plants can be relatively easily measured by combustion, a process that effectively reverses photosynthesis:

$$C_6H_{12}O_6 \text{(organic matter)} + \text{oxygen} \rightarrow$$
$$\text{carbon dioxide} + \text{water} (+ \text{energy})$$

The organic matter in this case will be mostly in the form of *carbohydrates*, which are organic molecules made up primarily of the elements *carbon*, *hydrogen*, and oxygen. Carbohydrates include sugars and other more complex molecules that use sugars as building blocks. Organic matter can also include other combustible molecules that contain lesser amounts of oxygen, such as fats. Food laboratories combust foods to measure their calorie content directly. The energy released is chiefly in the form of heat rather than light and is therefore measured by an increase in temperature. The energy contained in one food calorie is sufficient to heat one kilogram of water (equal to approximately one liter) by one degree Celsius. Put differently, you could boil a liter of water with the calories found in a cinnamon roll.

The energy that land plants store generally represents less than 1 percent of the sunlight that strikes their leaves. The other 99+ percent is lost by reflection of visible light, the reradiation of invisible heat, and the conversion of water into water vapor. This is terrible performance for a solar collector. This efficiency calculation also does not include any of the energy necessary actually to plant, cultivate, and harvest the plants or to process them prior to energy conversion. The latter energy investments can be considerable, especially when crop energy is recovered through fermentation and distillation of alcohol fuels such as ethanol.

In fact, entomologist David Pimentel and petroleum engineer Tadeusz Patzek calculated that the production of ethanol from corn actually results in a net *loss* of energy, implying that biofuels are not really a useful energy source at all. This conclusion was quickly attacked by other researchers, who have argued that Pimentel and Patzek failed to include the energy value of biomass by-products in their calculation. When these by-products, known as distiller's grains, are included, the calculated ratio of useful energy obtained from biofuels to input energy improves to something in the range of 1.3 to 1.6 to 1. Put differently, corn ethanol provides a 30 percent to 60 percent rate of return on the energy invested in obtaining it. On the positive side such ratios mean that the energy presently available from fossil fuels could be increased proportionately, simply by investing it in corn ethanol production. This new increment of energy would, in principle be gained without the need to emit additional CO_2 to the atmosphere.

A 30 percent rate of return in the financial markets is the stuff dreams are made of, and compounded over time would generate immense wealth. Unfortunately the same is not true for energy production systems. Early oil fields provided energy output-to-input ratios on the order of 100:1, back in the days when "gusher" was a common descriptor of a successful well. A 100:1 ratio equates to a 10,000 percent rate of return. Even the *least* efficient oil fields still return energy at ratios on the order of 5:1, equal to a 400 percent rate of return. The comparatively low rates of energy return afforded by corn ethanol might nonetheless be acceptable, if no other (nonenergy) costs were incurred. Unfortunately this is not the case.

Biofuels and Land Use

The intrinsically low efficiency of plants as solar collectors necessitates the cultivation of large areas of land, if biofuels are to make a substantial contribution to transportation fuels. This basic fact is inescapable, but the actual energy return ratios for various biofuel crops and the potential for competition between energy versus food crops are hotly debated. As a reference point in this debate, it is useful to consider the proportion of U.S. agriculture currently devoted to ethanol, which is produced almost entirely from corn. According to the U.S. Department of Agriculture, nearly 30 percent of all cultivated land was planted in corn in 2013. A record crop was harvested, 35 percent of which was used in making ethanol (this percentage was even higher in the previous year because of a smaller crop). In net terms, about 10% of U.S. cropland was used for ethanol production.

The ethanol was blended with gasoline at an average ratio of about 10 percent of the finished product. A tenfold increase in ethanol production would thus be required to replace gasoline completely. To do so would require the entire area of cropland available in the United States. This estimate assumes however that ethanol and gasoline contained equal amounts of energy. In truth they do not; ethanol provides approximately two-thirds as much energy per liter as gasoline. For corn ethanol to entirely replace gasoline would therefore require approximately 50 percent more cropland than presently exists. Even this calculation is perhaps wildly optimistic, since it assumes that corn can be made to grow productively in places where it presently does not. It is clear that a complete switchover from gasoline to corn ethanol would be a physical impossibility in the United States, even if 100 percent of the food supply were imported.

Fortunately, not all ethanol crops are created equal. Sugarcane, for example, stands out among land plants, able to capture nearly 2 percent of incident sunlight. The energy return ratio for sugarcane ethanol has been estimated to be roughly three times that of corn ethanol. Sugar cane won't grow just anywhere, however; it favors warm tropical climates with plenty of rain. Equator-straddling Brazil produces nearly twice as much sugarcane as any other country. Brazil, India, and China together produce about two-thirds of the world supply. Sugarcane accounted for only 2.8 percent of all Brazilian cropland in 2008, but 55 percent of this crop was used for ethanol production. Of the remainder, 44 percent went into sugar production, and 1 percent was used to make alcoholic beverages!

Brazil mandates that gasoline contain 20–25 percent ethanol, in contrast to the United States, where ethanol is limited to 10 percent in E10 fuels. The total demand for gasoline in Brazil could be satisfied using only about 15 percent of the presently available cropland. This would in principle leave plenty of land available to grow food, and to indulge the national taste for Caipirinhas (an alcoholic drink made from fermented sugarcane).

Large-scale sugarcane production unfortunately is not practical at higher latitudes; for example, the United States produces less than one-twentieth as much sugarcane as Brazil, most of it in Florida. However, in recent years a third option has entered the debate: "cellulosic" ethanol. Cellulose, or more properly lignocellulose, is the basic structural building block of most land plants. Cellulose contains literally explosive amounts of energy: It is the key ingredient used in manufacturing modern gunpowder (nitrocellulose). It was also used to manufacture early photographic films, before the highly flammable nature of those film stocks led to their replacement by more stable plastics made from petroleum. It is widely agreed that the energy return ratios on cellulosic biofuels could be much higher than for corn ethanol, and would perhaps rival those for sugarcane.

Give Yeast a Chance: Fermentation and Biofuels

So why don't our cars already run on "grassahol"? To answer this question it helps to examine the history of intoxication. Humans have been producing ethanol from grains and sugars since the Stone Age, and it is clear that our ancestors enjoyed their drink. For example, archaeologists have found the preserved remnants of wine in jars dating back six thousand years, and beer jars dating to perhaps ten thousand years ago. The original technology of ethanol production was likely discovered rather than invented, as a result of accidental food spoilage. Ethanol is produced as a toxic by-product by certain yeast (microorganisms), via a process known as fermentation:

$$C_6H_{12}O_6 \text{ (Sugar)} \rightarrow 2C_2H_5OH \text{ (ethanol)} + 2CO_2 \text{ (carbon dioxide)} + \text{heat}$$

Unlike simple combustion, fermentation is an anaerobic process, meaning it takes place without the addition of oxygen. Yeast cannot completely convert sugar all the way to water + carbon dioxide, however, as could be done by combustion. Instead they only partly reverse the process of photosynthesis, producing ethanol instead of water.

Some of the energy contained in the sugar is used up by the yeast or converted to heat and therefore lost, but most remains as chemical bonds in the ethanol. As an extra bonus, carbon dioxide release adds the bubbles to beer or Champagne.

Like people, yeast can only handle so much alcohol. A blood alcohol content of 0.50 percent would kill any human, but yeast can handle ten times this much or more in their water supply. Depending on the type of yeast, they call it quits when the alcohol content of the water in which they live reaches from 5 to 16 percent. Such concentrations make for a reasonably strong drink but are nowhere near high enough to support combustion. To obtain a product that can put hair on your chest or horsepower in your car requires that ethanol be separated from water by distillation. Distillation exploits the fact that ethanol boils at 78.1°C whereas water boils at 100°C. The mixture is heated above the boiling point for ethanol, but below that of water. The resultant vapor is then collected and condensed back to a liquid.

Because of chemical interactions with ethanol, some water is carried along in the vapor, resulting in a distilled mixture that is at best only about 95 percent pure. This hydrous ethanol, the equivalent of 190 proof spirits, will readily support combustion. It will not mix with gasoline, however, requiring additional steps to produce anhydrous, 200 proof ethanol. Unfortunately distillation and the removal of the remaining water both cost energy. In the United States ethanol refineries the energy for distillation results principally from combustion of natural gas, but a variety of other sources (including ethanol itself) could be used. The amount of energy needed varies with the efficiency of the distillery, but it may represent on the order of half the total energy content of the produced ethanol.

Yeast are picky eaters, primarily interested in sugar molecules such as glucose, sucrose (cane sugar), or fructose (fruit sugar; see Figure 5.6). More complex carbohydrates called starches can also be used to make ethanol, provided the yeast have some help breaking them down into their constituent sugars. To do so requires an intervention by enzymes, which are compounds that serve to promote specific chemical reactions. For example, barley releases the enzyme amylase when it is germinated in water, in a process known as malting. Amylase in turn helps to liberate the fermentable sugars needed to make beer or whiskey. This is all ancient "off the shelf" technology, the exact origin of which has been long forgotten.

Liquor historians and observant laypersons will note that we have no tradition of making alcoholic beverages from cornstalks or

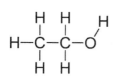

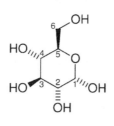

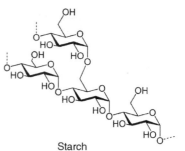

Ethanol   Glucose   Starch

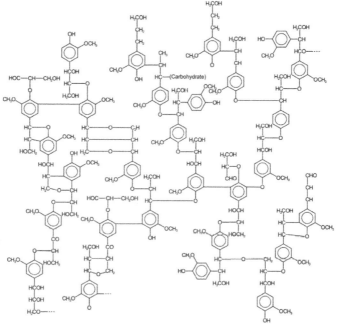

Lignin

Figure 5.6. Schematic structures of ethanol, glucose, starch, and lignin (Wiki Commons).

grass clippings. Like starch, cellulose consists of long chains of constituent sugar molecules, but breaking down these chains requires enzymes that are less readily available. Cellulose in woody plants is also tightly bound with lignin, a tough and complex natural polymer that does not yield to yeast. Complicating matters further, some of the sugars contained in cellulose are not readily fermentable by traditionally employed yeast strains. Specialist strains must therefore be brought in for the job, or else be genetically engineered. None of these challenges appears to be insurmountable, but industrial-scale ethanol production from lignocellulose still lies largely in the future.

Cutting Out the Middle Man: Biofuels Direct from Plants

What if we could sidestep the need for yeast entirely, and with it the energy wasted on the distillation of alcohol? Why can't plants make liquid fuel, unassisted? Actually they can. The ancient Greek word for oil was synonymous with olive, and the chemical structure of olive and other vegetable oils is a close cousin to that of petroleum. All these substances are dominated by lipids, which may be broadly defined as organic compounds that are insoluble in water (the word "lipid" is derived from the Greek word for fat). Plants synthesize lipids as part of their protective cell wall membranes and for energy storage. Oil may be easily extracted from various seeds and legumes, for example, sunflower seeds, rapeseed, peanuts, and soybeans. The German inventor Rudolf Diesel demonstrated in 1893 that the engine that now bears his name could run on peanut oil alone.

Diesel fuel derived from petroleum consists of various lipid compounds that generally contain between approximately ten and fifteen carbon atoms each. Such compounds occur naturally in petroleum and may be separated by distillation at relatively low temperatures. Vegetable oils also contain lipid compounds with the appropriate number of carbon atoms, in the form of fatty acids. However, these lipids mostly occur as part of more complex molecules called triglycerides, which consist of three separate fatty acid chains that are attached to a single alcohol molecule (glycerol). To make true biodiesel the lipids must be liberated from the triglycerides; this is achieved by treatment with an alcohol such as methanol or ethanol. In a tongue-twisting reaction known as transesterification, glycerol is replaced by a single ethanol molecule attached to each of the three lipids (Figure 5.7). The lipids are now suitable for fuel use, and glycerol is sent to find other employment, for example, as the featured ingredient in the explosive nitroglycerine. While not purely a virgin fuel, biodiesel nonetheless requires less energy input for processing than does ethanol.

Calculation of the overall energy balance of biodiesel is less clear, in part because the oil-producing seeds or legumes represent a relatively small part of the overall biomass. It appears likely that the energy return on energy invested is significantly better than for corn ethanol, but perhaps not as good as that of sugarcane or cellulosic ethanol. A recent study published by researchers from the University of Idaho and the U.S. Department of Agriculture concluded that the

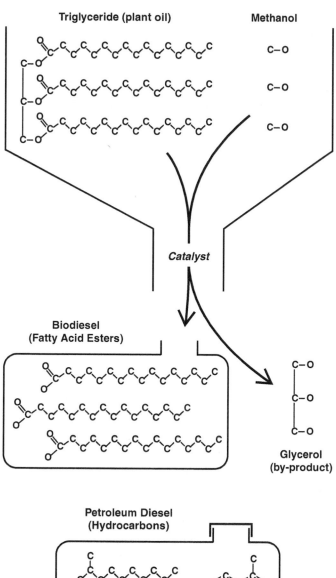

Figure 5.7. Simplified summary of transesterification.

Triglycerides (plant oil) and methanol are reacted in the presence of a catalyst to produce biodiesel and glycerol. Biodiesel has chemical structures similar to those found in petroleum diesel, shown here for comparison. C = carbon atom, O = oxygen atom (hydrogen atoms omitted for clarity).

average energy return ratio for soybean biodiesel is 2.55:1 (Pradhan et al., 2008).

Biodiesel made from recycled animal fat or vegetable oil might boost the returns still further, but only on a very limited scale. The U.S. Department of Energy's Idaho National Engineering and Environmental Laboratory made headlines in 1999 by announcing a method for making biodiesel from the oil left over from cooking french-fried potatoes. While this strategy commendably reduces waste, the available volume of used fry oil pales in comparison to our annual petroleum demand. Perhaps if the world's population converted to a diet of nothing but french-fried potatoes and other fried foods we could all ride in biodiesel-fuels cars, but the negative health effects of such a diet would no doubt outweigh its energy benefits.

Back to the Sea

Perhaps nature knows best. Most of the naturally occurring oil within the Earth was generated from algae and other aquatic microorganisms rather than from land plants. Algae, while more primitive than land plants, possess certain built-in advantages with respect to the conversion of solar radiation into liquid fuel. Because they have no need for rigid, gravity-defying structures, algae generally do not contain lignocellulose. They do contain a very high proportion of lipid compounds that are readily converted into liquid fuel. Algae grow quickly and make relatively efficient use of sunlight, permitting biofuel yield rates that are perhaps thirty times higher than those for land plants covering the same area. Microalgae need no soil, and therefore no productive agricultural land that could otherwise have been used to grow food. They do need water, but are not picky about its salinity or cleanliness. Seawater will do, or even sewage. Algae also need nutrients such as nitrogen and phosphorus, which represent pollutants when overabundant in freshwater supplies. Biofuel production from algae might therefore do double duty by also improving water quality.

As will be detailed in Chapter 7, the Earth possesses natural (but extremely inefficient) systems for collecting algal biomass and converting it into liquid fuel. Unfortunately humans do not, at least not yet. Existing large-scale biofuel production from corn, sugarcane, soybeans, and other plants required relatively little research or development. For the most part these fuels have been obtained by adaptation of known technology and existing infrastructure, which had already been developed for food production. No such infrastructure exists for algae; currently humans are about as likely to put algae on their faces (as a cosmetic) as they are to put it into their stomachs. It is therefore a relatively expensive specialty product that needs significant research and even more significant infrastructure investment before it can become a major fuel source. Much of this work is already under way, focused on finding or engineering the most appropriate microalgae strains for biofuels, maximizing the production of lipid biomass, improving cultivation efficiency, and other problems.

Conclusions

Plants might be thought of as cheap, mass-produced solar energy collectors. They suffer from substantially lower efficiencies than photovoltaic cells, but their production costs are substantially lower as well. As a result they appear capable of competing on a cost basis with other forms of solar energy collection. As with other solar energy systems, however, the energy captured by plants is also limited by the generally diffuse nature of sunlight. Large amounts of land surface area are unavoidably required to collect meaningful amounts of bioenergy.

Plants have been keeping us warm, well fed, and intoxicated for millennia. The first human use of bioenergy, via direct combustion of plant matter, predates recorded history. The systematic cultivation of grains and other agricultural food stocks was an important cornerstone of early civilization. Advances in farming continue to support more the more than 7 billion people living on Earth today. The basic technologies that enable ethanol production from simple carbohydrates and oil extraction from seeds and legumes also date from near the beginning of human civilization.

Adapting these ancient technologies to the enormous scale of modern energy usage presents certain challenges. For example, the net gain of energy from corn ethanol is modest, compared to the amount of energy that must be invested to obtain it. Such problems do not appear insurmountable, however. Improved technologies such as ethanol production from lignocellulose appear within reach and promise better energy yields. A more troubling problem is the potential competition between biofuels and food, for the Earth's limited supply of productive farmland. Biofuels can only be considered feasible if the world's total agricultural capacity exceeds that needed to sustain its population.

For More Information

Buol, S. W., Southard, R. J., Graham, R. C., and McDaniel, P. A., 2003, *Soil Genesis and Classification*, 5th ed.: Ames, Iowa, Blackwell Publishing Professional, 494 p.

Cleveland C. J., 2005, Net energy from the extraction of oil and gas in the United States: *Energy* v. 30, pp. 769–782.

Farrell, A. E., Peving, R. J., Turner, B. T., Jones, A. D., O'Hare, M., and Kammen, D. M., 2006, Ethanol can contribute to energy and environmental goals: *Science*, v. 311, p. 506–508.

Hall, D. O., and Rao, K., 1999, *Photosynthesis*: Cambridge, Cambridge University Press, 214 p.

Hill, J., Nelson, E., Tilman, X., Polasky, S., and Tiffany, D., 2006, Environmental, economic, and energetic costs and benefits of biodiesel and ethanol biofuels: *Proceedings of the National Academy of Science*, v. 103, p. 11,206–11,210.

McGovern, P. E., and Mondavi, R. G., 2007, *Ancient Wine: The Search for the Origins of Viniculture*: Princeton, NJ, Princeton University Press, 400 pages.

Milbrandt. A., 2005, A Geographic Perspective on the Current Biomass Resource Availability in the United States: United States National Renewable Energy Laboratory Technical Report TP-560–39181, 70 p.

Montgomery, D. R., 2006, *Dirt: The Erosion of Civilizations*: Berkeley, University of California Press, 296 p.

Perlack, R. D., Wright, L. L., Turhollow, A. F., Graham, R.L., Stokes, B. J., and Erbach, D. C., 2005, Biomass as a Feedstock for a Bioenergy and Bioproducts Industry: The Technical Feasibility of a Billion-Ton Annual Supply: United States Department of Energy Oak Ridge National Laboratory, 78 p.

Pfeffer, M., Wukovits, W., Beckmann, G., and Friedl, A., 2007, Analysis and decrease of the energy demand of bioethanol-production by process integration: *Applied Thermal Engineering*, v. 27, p. 2657–2664.

Pimental, D., 2003, Ethanol fuels: Energy balance, economics, and environmental impacts are negative: *Natural Resources Research*, v. 12, p. 127–134.

Pimentel, D., and Patzek, T. W., 2005, Ethanol production using corn, switchgrass, and wood; biodiesel production using soybean and sunflower: *Natural Resources Research*, v. 14, p. 65–76.

Pradhan A., Shrestha D. S., Van Gerpen J., and Duffield J., 2008, The energy balance of soybean oil biodiesel production: A review of past studies. *Transactions of the American Society of Agricultural and Biological Engineers* v. 51, p. 185–194.

Schmer M. R., Vogel K. P., Mitchell R. B., and Perrin R. K. 2008. Net energy of cellulosic ethanol from switchgrass. *Proceedings of the National Academy of Sciences of the United States of America* 105: 464–469.

Soil Survey Staff, 1999, *Soil Taxonomy: A Basic System of Soil Classification for Making and Interpreting Soil Surveys*, 2nd ed.: United States Department of Agriculture Natural Resources Conservation Service, *Agriculture Handbook* 436, 871 p.

6

Fossil Farming: The Geologic
Underpinnings of Biofuels

> We may be quite sure that among plants, as well as among ani-
> mals, there is a limit to improvement, though we do not exactly
> know where it is.
>
> *Thomas Robert Malthus, 1798*

In 1902 the English author W. W. Jacobs penned a short story entitled
"The Monkey's Paw," about a shriveled artifact with the magical power
to grant its owner three wishes. Unfortunately those wishes resulted
in unexpected and terrible consequences that far outweighed their
benefits, to the bitter regret of the wisher. The cautionary message of
this tale is that we should be very careful any time something valuable
is offered without apparent cost.

Biofuels promise to grant only two wishes rather than three,
but they are good ones: reduced greenhouse gas emissions and energy
renewability. However, does the granting of these wishes carry a dark
side that might outweigh their apparent benefit? The most obvious
concern is the potential for biofuels to compete with food production.
Part of the reason we can seriously consider biofuels at all is the "Green
Revolution," a dramatic increase in global farm output that occurred
during the latter half of the 20th century. According to data published
by the U.N. Food and Agriculture Organization, world production of
rice, wheat, and corn roughly tripled between 1961 and 2007. Some
individual crops or countries experienced even larger increases. For
example, average corn yields in the United States increased roughly
fivefold between 1950 and 2000, after having been essentially flat during the first half of the century (Figure 6.1). Total wheat production in
China increased more than sixfold during 1961–2008.

This agricultural bounty offers the hope that we can burn our
cake, and eat it too. However, increased crop yields carry their own

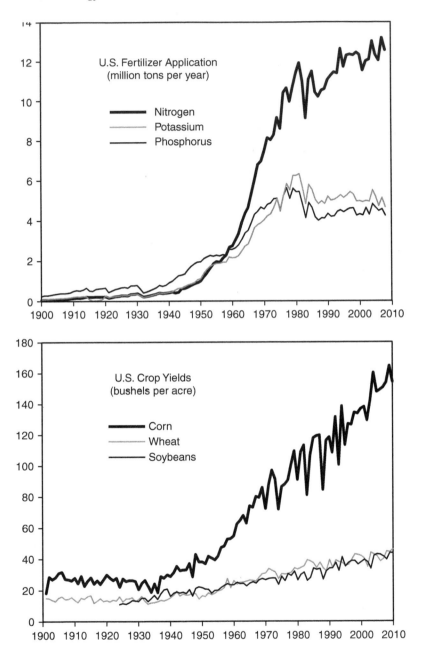

Figure 6.1. Historical U.S. fertilizer application and crop yields (data from U.S. Department of Agriculture).

price, in the form of nonrenewable Earth resources needed to support them. Leading the list are fossil fuels, used for everything from powering tractors to producing nitrogen fertilizer. Other vital fertilizers are also derived from limited geologic deposits. The soil needed to grow biofuel crops is itself a geologic resource, which is rapidly being depleted. Even the freshwater required to sustain high rates of productivity is a finite and threatened resource.

The Growth of Agriculture

The Green Revolution was only the latest chapter in a long history of increasing global food production. Farm output has been steadily increasing for centuries, in part through the gradual expansion of agriculture into previously undeveloped lands. This expansion did not occur at random, however. It closely followed natural patterns of soil fertility, which in turn are governed by a fortuitous combination of geology and climate.

At the dawn of the 18th century the major agricultural regions, as measured by fraction of land area used to grow crops, comprised eastern China, India, Europe, Russia, and parts of sub-Saharan Africa (Figure 6.2). Agriculture in China focused mostly on its eastern river plains and deltas, which are dominated by the Yellow and Yangtze Rivers (Figure 5.3). Recurrent floods are a defining feature of Chinese history, simultaneously responsible both for widespread destruction and for soil replenishment. The Indo-Gangetic river plain in northern India experienced a similar early history. The most intensive agriculture in Africa developed within what is now Nigeria, based on a band of forest soils that lies in the "Goldilocks zone" south of the Sahara desert and northwest of the central African rain forest. Agriculture was also well established throughout Europe in 1700 and based on a complex mosaic of riverine, forest, and grassland soils.

Crop cultivation had greatly intensified by 1850 on the eastern China plains. Agricultural intensification also continued in northern India and on the Deccan Plateau. This plateau consists of an enormous area of volcanic lava, which erupted about the same time that the dinosaurs disappeared (~66 million years ago). Exposure to the elements partly converted the surfaces of these lava flows to clay minerals, which shrink and swell depending on how much water is available. A relatively fertile soil characterized by deep vertical cracks developed

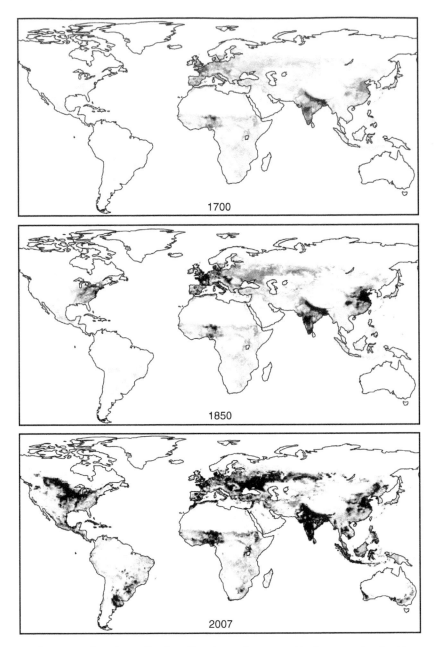

Figure 6.2. Fraction of land surface devoted to crops (darker = greater fraction; see Ramankutty and Foley, 1999, for methodology used).

(Based on data originally developed by the Sustainability and Global Environment program at University of Wisconsin-Madison and subsequently updated by the Land Use and Environmental Change Program, McGill University. Data downloaded from http://www.geog.mcgill.ca/~nramankutty/Datasets/Datasets.html) (Navin Ramankutty and Amy Kimball) as a result. Agriculture had also accelerated in western Europe and on the Eurasian steppe by this time and was well-established in the eastern United States. The Industrial Revolution had begun but was still new, and as a consequence farming still relied heavily on animal or human labor. World population was only about one-sixth of what it is today.

By the end of the 20th century virtually no naturally fertile soil region was left untouched. By this time the highly fertile, glacially derived soils of North America had been heavily developed for corn and wheat. Similar regions in Eurasia had also experienced agricultural growth, particularly in the Ukraine (the area north of the Black Sea). The Patagonian grasslands of Argentina had also been developed. A new pattern of increased development of marginally fertile soils in humid tropical areas had also emerged, intensifying crop growth in Central America, Southeast Asia, and parts of Brazil. Sustained farming in such areas typically requires modification of naturally acidic soils with lime and substantial addition of nutrients through fertilizers. This approach has been quite successful locally, for example, in the Cerrado region of Brazil.

Fertilizers have been used to boost the productivity of even naturally fertile soils for millennia, but until recently most of the fertilizers used were organic in the truest sense. They emanated from organisms, in the form of animal wastes or composted plant remains. One of the highest-quality natural sources of nitrogen was the accumulated droppings of sea birds and bats, mined from coastal caves in Chile and Peru. In the 19th century these deposits also acquired considerable strategic importance due to the use of extracted potassium nitrate for making gunpowder. Today, however, organic fertilizers make a relatively small contribution; for example, they represent only about 3 percent of the fertilizers applied to U.S. crops. The historically high yields of 20th century agriculture have instead relied on large-scale production of fertilizers obtained from nonrenewable geologic sources. As with fossil fuels, these materials are being extracted from the Earth far more quickly than they are being replenished.

Out of Thin Air: Inorganic Nitrogen Fertilizers

One of the more significant technological innovations leading to the Green Revolution occurred early in the 20th century, when the German chemist Fritz Haber demonstrated a process for making artificial nitrogen fertilizer. This process was further refined by Carl Bosch, of the German chemical company BASF. The Haber-Bosch process has sometimes been called the most significant invention of the 20th century, and both men received the Nobel Prize for their efforts. Ironically, Haber is also well known for his less beneficial contributions to the development of poisonous gases for use in warfare.

The Haber-Bosch process combined atmospheric nitrogen, which consists of pairs of strongly bonded nitrogen atoms, with hydrogen to make ammonia:

$$N_2 + 3H_2 > 2NH_3$$

Ammonia can be used directly as a fertilizer or as a feedstock to make other nitrogen fertilizers. The hydrogen required by the Haber-Bosch process today is derived mostly from methane, from natural gas. In addition to using natural gas as a feedstock, the reaction itself is highly energy-intensive, requiring pressures 150–250 times greater than atmospheric pressure and temperatures in the range of 300°C to 550°C.

The dramatic late 20th century increases in U.S. corn yield were accomplished in large part through the increased application of nitrogen fertilizer, produced by the Haber-Bosch process. The production of inorganic nitrogen fertilizer currently accounts for about 2 percent of U.S. natural gas production, a surprisingly large number considering that the United States accounts for nearly 20 percent of world natural gas production. Put differently, the natural gas used for heating and cooling by an average U.S. household over the course of one year would produce enough nitrogen to fertilize about thirty-five acres of corn. According to U.S. Department of Agriculture estimates for 2009, fertilizers (including nitrogen) represented approximately 25 percent of total corn production costs on average.

From Ancient Seas: Phosphorus and Potassium Fertilizers

Intensive crop growth also depletes two other major nutrients, phosphorus and potassium. These elements are abundant in the ocean, where they have accumulated as a consequence of the dissolution of rocks on the continents. Spraying fields with raw seawater will do nothing to boost their fertility, however, since the salt will kill most crops. What is needed is a way to concentrate phosphorus and potassium from seawater in a form that excludes the salt. Luckily nature has already done this for us, producing sedimentary rocks that we can mine for fertilizer.

The amount of phosphorus contained in seawater is very small, on the order of one part per million. It is relatively less abundant in surface waters, because of the demands of photosynthetic algae and other organisms that grow there, and more abundant in deeper waters. To form useful deposits these deep waters must therefore upwell to the shallower depths in the ocean, a process that is controlled by interactions between major ocean currents and coastlines (Figure 6.3). Upwelling tends to be particularly strong near the western coastlines of continents. Once it has risen from the deep ocean, phosphorus can precipitate within sediments (particularly mud and silt) that have been deposited in relatively shallow seas.

The initial phosphorus deposits formed in this way are usually still too lean to exploit, however. Winnowing of these deposits by ocean currents and waves removes mud and leaves behind naturally concentrated phosphatic minerals, fish bones, and teeth. Because of the requirement for several favorable factors to coincide, major phosphate deposits (called phosphorites) are geologic rarities that require millions of years to form (Figure 6.3). Most of the world's currently known deposits occurred on the northern to northwestern edge of Africa and the Middle East during the Cretaceous period, within the now-extinct Tethys Ocean. Morocco holds the majority of this resource. Lesser deposits are found in a variety of other countries, including China, the United States, and South Africa.

High-grade phosphorite deposits clearly will not last forever, but their true magnitude has been somewhat controversial. The U.S. Geological Survey in 2010 estimated global reserves of about 16 billion metric tons, of which one-quarter resided in Morocco. Assuming annual consumption equal to that in 2009 (158 million tons), those reserves would be expected to last only about one hundred years. Should we be worried therefore about running short? Perhaps. A number of people have suggested this idea; for example, Dana Cordell and coauthors published a paper in 2009 predicting that phosphorus production will peak around 2030 or shortly thereafter, then begin to decline. If correct, this prediction would imply dire consequences; there is no substitute for phosphorus in food production and we simply cannot survive without it.

However, reliable data on the true magnitude of phosphorite reserves are elusive. In 2010 the International Fertilizer Development Center (IFDC) proposed that the world's supply was actually about four times higher than earlier estimated, with most of the increase reported in Morocco. The U.S. Geological Survey soon quadrupled its

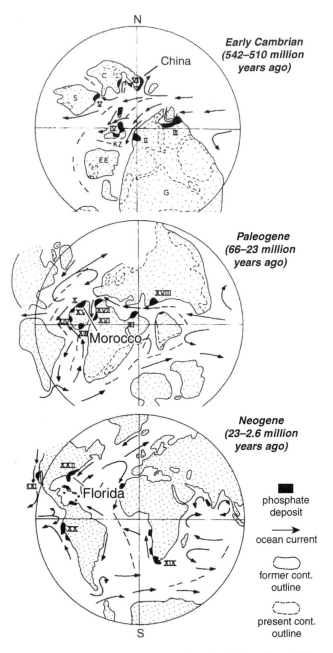

Figure 6.3. Deposition of major phosphate deposits during three different geologic periods.

Note that such deposits are related to fortuitous configurations of the continents with respect to ocean currents (modified from Glenn et al., 1994).

own estimate of global reserves to 65 billion tons, an amount equal to about a three hundred year supply at current production rates. Morocco is now estimated to hold two-thirds of the world's phosphorite reserves.

Potassium is much more abundant in seawater, and in contrast to phosphorus anyone can easily extract it. All it takes are a bucket and some patience. Go to the seashore, fill the bucket with water, place it in the hot sun, grab a beer, and wait. As the water dries up, various minerals will crystallize in a predictable order. First is calcite ($CaCO_3$), the main mineral in limestone. Gypsum ($CaSO_4$), the main ingredient in drywall and plaster of Paris, will begin to precipitate when about 80–90 percent of the original water in the bucket has evaporated. Next is halite (sodium chloride, or table salt), which precipitates when about 86–94 percent of the original water is gone. Only when the bucket is nearly dry (>94 percent of the water has evaporated) will potassium- and magnesium-bearing minerals make their appearance. This same basic approach to obtaining potassium could, in principle, be applied at the industrial scale, using large evaporation ponds constructed adjacent to the ocean. This process is laborious and time-consuming, though, and produces far more halite than potassium salts.

Most of the world's potassium production results from mining of geologic deposits of minerals formed by past evaporation. In contrast to the bucket experiment, the Earth has on occasion managed to create thick layers of potassium evaporite (called potash), without generating the expected large amounts of halite. This trick is not easy even when millions of years are available to attempt it and requires an unusual combination of circumstances. In nature, a mostly closed inland sea takes the place of a bucket. This inland sea must be located where sunlight is strong and evaporation rates are high, for example, at or near the equator. The modern Mediterranean Sea fits the bill, since it is connected to the ocean only by the Gibraltar Straits and enjoys a warm, dry climate. In fact it did almost dry up about 5 million years ago, but because the Mediterranean basin is quite deep (up to 5 km) its connection to the ocean was quickly reestablished. Today the Mediterranean is only slightly saltier than the rest of the ocean.

Going further back in time, 400 million years ago the province of Saskatchewan in southern Canada lay near the equator and was home to a much shallower, isolated sea that flooded part of the continent. This sea was only tenuously connected to the ocean, and much of the time it was nearly dry. Potassium salts could therefore crystallize near its more inland extremities. Just enough new seawater flowed in periodically to resupply it with potassium. A delicate balance between inflow and evaporation was maintained for millions of years, during which thick deposits of potassium-bearing minerals accumulated. As a result of this geologically rare confluence of circumstances, nearly half of the world's potash reserves are found in Saskatchewan. Russia places a distant second. In 2012 the U.S. Geological Survey estimated the world's known reserves at 9.5 billion tons, versus annual production of only 34 million tons. Although still finite, these reserves should therefore last several centuries at least. Even if they run out, there will be plenty of potassium remaining in the ocean.

The secondary nutrients calcium, magnesium, and sulfur are extremely common, and their future availability is not in doubt. The picture is more complicated for micronutrients, because most of these are mined primarily for industrial uses rather than for agriculture. The known reserves are generally equal to forty to sixty years of supply at current production rates. Such reserves estimates are misleading, however, because they reflect the industrial value of those elements, rather than their value as micronutrients.

Washing Away: Agriculture and Soil Erosion

Rome wasn't built in a day, and neither were the soils needed to produce food or biofuels. Natural soil formation is a slow process that is measured in hundreds to thousands of years, and sometimes more. For example, the rich soils of the U.S. Corn Belt formed largely on glacial sediments, which were deposited at least as twelve thousand years ago. As time passed these soils evolved in response to the changing climate and ecology. Minerals in the glacial sediment were gradually dissolved, releasing nutrients such as Ca, Mg, and K to be used by plants. Countless generations of earthworms, insects, and rodents busily churned the soil, thoroughly rearranging its original structure and texture. Plant roots and other organic matter gradually accumulated until balanced by natural decomposition, producing the dark colors associated with rich topsoil in mollisols and alfisols.

American blacksmith John Deere is widely credited with helping to open the American prairie to agriculture in 1837, by inventing a steel plow that could break tough and sticky sod to expose the rich soil beneath. This, and similar technological advances that occurred throughout the Industrial Revolution, laid bare some of the world's richest soils for the first time. Soils, which are end products of rock weathering, ironically began to experience unprecedented rates of weathering themselves. At its simplest level this was expressed as soil erosion. Whereas natural root systems tend to bind soil and protect it from disturbance, tilling exposes it directly to wind and water. The potential destructive effects of wind were vividly illustrated during the Dust Bowl years of the 1930s, when a large amount of topsoil from the southern Great Plains of the United States blew east in great billowing clouds. Many of the local inhabitants were driven west to resettle in California as a result.

Water attacks the landscape even more violently. It has often been said of the Missouri River that it is "too thick to drink, too thin to plow," because of the large amount of suspended sediment it carries. Most of this sediment derives from high elevations in the northern Rocky Mountains, where steep slopes and winter ice conspire to dislodge it. Erosion also occurs at lower elevations, through processes such as raindrop impact and the flow of water across exposed soils, but natural erosion of soil-covered flatlands is generally much slower than in the mountains. Tilling the soil and replacing natural vegetation with row crops can dramatically accelerate these normally placid processes, by destroying root networks that bind the soil and by removing overlying plant matter that shields it. The degree of acceleration varies widely with geography and agricultural practices, but there is no doubt that soil is now being lost at geologically unprecedented rates. David Montgomery of the University of Seattle estimated in 2007 that on average, soil loss rates associated with agriculture exceed natural soil erosion by a factor of 10 to 100.

Comparing average modern soil loss rates to geological rates of erosion, Bruce Wilkinson and Brandon McElroy (Syracuse University and University of Texas, Austin, respectively) recently concluded that humans are now by far the dominant agent of landscape modification on Earth. In fact, sediment removal from soil has been so rapid that it has apparently outstripped the capacity of rivers to carry it away. It therefore piles up close to its source, within gully bottoms and on floodplains, until such time as onward transportation to the sea can be arranged. Called "postsettlement alluvium" in reference to its historically recent origin, this sediment can be easily recognized where it buries more fertile, native soils.

Rapid in this case is relative, and if there is any bright side to the preceding scenario it is that overpopulation will likely overtake us long before soil erosion does. The average cropland erosion rate used by Wilkinson and McElroy is less than a millimeter per year, and soil thickness is often a meter or more within the Corn Belt, implying that soil depletion in the U.S. heartland is generally more of a millennial problem than an immediate crisis. Local erosion rates can be far worse, of course, but erosion can also be ameliorated through improved agricultural practices such as no-till farming. Wilkinson and McElroy conclude that world population is increasing at approximately twenty-five times the anticipated rates of human-induced soil loss, a conclusion that if true suggests that the latter problem could be moot.

Raising the Dead: CO_2 Emission from Soil Carbon

Soils contain huge amounts of organic carbon, accumulated from the decay of dead plants. Because it accumulated over very long periods, this carbon can reasonably be thought of as fossil carbon. Carbon buried 1,000 years ago in soil is really no different from carbon buried 100 million years ago in coal; a corpse is a corpse regardless of its antiquity. It is difficult to precisely measure the amount of carbon buried in soil; most estimates put it at around twice the amount of carbon presently stored in the atmosphere. Release of soil carbon to the atmosphere as either CO_2 or CH_4 (methane) could therefore have an identical effect on global temperatures to the burning fossil fuels.

Plowing turns over the surface layer of soil and exposes it to the atmosphere, thereby exhuming some of its buried carbon. For long-established farm fields the release of soil carbon by plowing may be balanced by new burial of organic matter during the yearly crop cycle. If so, then the net impact on the atmosphere should be nil. However, the situation is very different when native forest or grassland soils are plowed for the first time. The release of fossil carbon from newly plowed fields exceeds the amount of carbon that can be reburied as crop residue. Such land use change therefore does cause a net addition to greenhouse gases to the atmosphere. In fact, soil carbon released by agricultural expansion appears to have been the largest source of anthropogenic greenhouse gas prior to the mid-20th century, when the burning of fossil fuels took over that dubious distinction (Figure 6.4).

Timothy Searchinger of Princeton University and colleagues argued in a 2008 report that increased conversion of forest and grasslands to croplands is a likely response to rising demand and prices for biofuels. They estimated that 25 percent of the soil carbon originally present in newly cleared land would be released, on average. Even if biofuels crops were grown on established agricultural land the end result would likely be the same, because new land would have to be

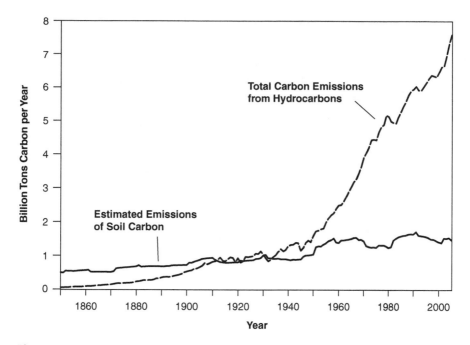

Figure 6.4. Estimated release of carbon from soils, compared to emissions that have resulted from burning hydrocarbon fuels (Data sources: Houghton, 2008; U.S. Department of Energy).

cleared to grow the food crops that had been displaced. Searchinger and his colleagues argued that rather than reduce carbon emissions, projected increases in the production of corn ethanol in the United States would actually cause CO_2 emissions to double within thirty years, compared to the alternative of continued gasoline use.

Searchinger and colleagues noted that although biofuels should eventually bring about decreased CO_2 emissions, they first have to compensate for the soil carbon released to the atmosphere by land use change. The payback period can be determined by comparing the amount of soil carbon that is released by plowing with the annual reduction in CO_2 emissions attributable to substituting biofuels for gasoline. The debt is not a trivial one; Searchinger and his colleagues estimated that paying it off might require as much as 167 years, in the case of land cleared to grow corn ethanol. The carbon debt for switchgrass appears a bit less intimidating but is still substantial, with an estimated payback period of 52 years. Searchinger and his colleagues concluded that the only way to obtain a true reduction in greenhouse gases over the short term would be through more extensive use of municipal or crop wastes, which presently are not used for either food or energy.

Green Gone Wild: The Downstream Consequences of Agriculture

Whatever is lost from the continents eventually finds its way to the ocean, carried there by rivers. This includes not only the tangible products of physical weathering (sand, silt, and mud), but also chemical elements released from the dissolution of rocks and soil. The latter serve as vital nutrients for algae, and by extension help to lay the foundation of the entire marine food chain. It is possible to get too much of a good thing, however. An overabundance of nutrients can lead to runaway algal productivity in coastal waters, a condition known as eutrophication.

The resultant green soup is relatively rich in oxygen near the water surface, because of mixing of oxygen from the atmosphere and photosynthesis. As algae die they sink, however, to be entombed in the deeper and darker waters below, where they are scavenged by aerobic bacteria. Aerobic in this case does not imply physical exertion (or the wearing of Spandex), but instead the necessity for these bacteria to consume oxygen to drive their metabolism. Increased productivity at the surface of the ocean therefore leads directly to a deficit of oxygen at depth, attributable to bacterial feasting. Bottom water hypoxia results if dissolved oxygen concentration drops below 0.2 ml/l. The river-fed surface waters are less saline and generally warmer than the waters below. The resultant density contrast inhibits mixing and seals the fate of marine organisms that cannot escape the fetid bottom. The result is commonly referred to as a "dead zone" where life either ceases or else is badly compromised. Shrimp, starfish, and other bottom dwellers, along with deep-swimming fish, are particularly hard hit.

Natural coastal eutrophication caused by river discharge is comparatively rare, because nutrients on land tend to be absorbed by natural vegetation, soils, and wetlands. Riverine concentrations are normally too low to cause widespread hypoxia. Overloading of rivers with nutrients from artificial sources can change this balance, however, and the post–World War II expansion in human population has made coastal eutrophication more common. A variety of sources have been implicated, including municipal sewage, oxidation of soil nitrogen due to plowing, destruction of wetlands, and accumulation of atmospheric nitrous oxides derived from burning of fossil fuels. It is widely agreed though that runoff of agricultural fertilizers (particularly nitrogen) is the main culprit.

Robert Dias and Rutger Rosenberg noted in a 2008 review, published in the journal *Science*, that dissolved oxygen began to decline in many areas about ten years after the increased use of nitrogen fertilizers, and that hypoxia first became widespread during the 1960s to 1970s as fertilizer application grew at a rapid rate. Some of the worst problem spots include Chesapeake Bay, the Gulf of Mexico, and the Baltic Sea; major degradation of fisheries has occurred in all of these areas. The Black Sea also began to experience hypoxia during the 1970s and the problem intensified through the 1980s, in response to nutrients carried to it by the Danube River. However, hypoxia completely disappeared between 1990 and 1995. This remediation did not occur in response to any environmental activism, but was a consequence of economic collapse in the former Soviet Union. As money to buy agricultural fertilizer and to support industrial livestock production disappeared, so did the flow of nitrogen from the Danube.

Hypoxia in the Gulf of Mexico has been blamed primarily on runoff from corn and soybean fields within the U.S. Corn Belt. The size of the Gulf of Mexico "dead zone" varies from year to year, depending on the amount and timing of discharge from the Mississippi and Atchafalaya Rivers; greater flow carries more nutrients. Gulf currents tend to spread the hypoxic zone westward, and at its largest it can span the entire Louisiana coast and reach eastern Texas (Figures 6.5 and 6.6). Over the past two decades the total flow of nitrogen to the Gulf has averaged about 1.5 million metric tons per year, with a little less than 1 million metric tons in the form of nitrate (NO_3^-). The remainder is dissolved organic nitrogen, particulate organic nitrogen, and ammonium (NH_4^+). Average annual nitrogen influx has roughly tripled since the 1960s, with most of the increase occurring in the 1970s to early 1980s. Since then it has remained relatively constant, as has total application of agricultural fertilizer over the same period.

The onset of hypoxia in the Gulf of Mexico was a relatively quiet environmental crisis; it was already well under way by the time it began to be recognized in the late 1980s. The first formal study to be published was that of R. Eugene Turner and Nancy Rabalais in 1991. Although initially a consequence of food production, the increased use of corn to produce ethanol gradually transformed the Gulf of Mexico "dead zone" into an energy problem.

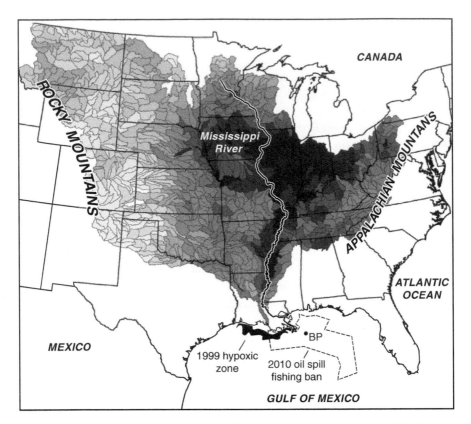

Figure 6.5. Onshore areas: total nitrogen delivery (incremental yields) from areas within the Mississippi River watershed (darker shading = higher mass of nitrogen delivered per unit area per year; values range from 0 to 5540.9 km/m²/yr).

Area of "fishing ban" is the fisheries closure boundary announced by the National Oceanographic and Atmospheric Association for May 25, 2010. BP = location of BP/Deepwater Horizon well (Map adapted from Roberts and Saad, U.S. Geological Survey, http://wi.water.usgs.gov/rna/9km30/index.html [downloaded 1/3/10]. Location of 1999 Gulf of Mexico hypoxic zone taken from Goolsby, 2000[b]).

Anticipated future increases in biofuel production, needed to meet current U.S. policy goals for renewable fuel, can only worsen the situation, according to a 2008 study published by Simon Donner of the University of British Columbia and Christopher Kucharik of the University of Wisconsin. These authors estimated that the U.S. goal of increasing renewable fuels to 15–36 billion gallons per year (about 11–26 percent of U.S. gasoline consumption) would increase delivery

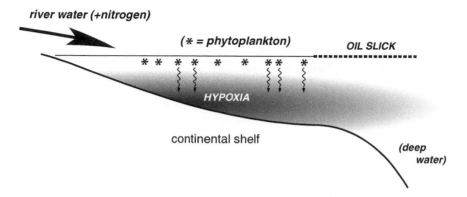

Figure 6.6. Schematic illustration of the "dead zone" in the Gulf of Mexico. Warm, freshwater from the Mississippi River flows over the denser marine waters of the continental shelf. Nitrogen and other nutrients carried by the river stimulate high rates of phytoplankton productivity near the surface. As they die they settle into deeper water, where they are consumed by aerobic bacteria, which rapidly exhaust the available oxygen to create an hypoxic zone. An oil slick, such as that formed by the 2010 spill, is shown for comparison. Note that because it is less dense than water, oil tends to form a very thin layer that can be spread out by winds and currents to cover a very large area of the ocean surface.

of dissolved inorganic nitrogen to the Gulf of Mexico by 10–36 percent. Great news for algae; not so good for shrimp fisheries. Potential solutions would have to be drastic to counter this impact. For example, the U.S. population could give up eating meat, freeing up existing corn production that presently is mostly fed to livestock. Alternatively the United States could embark on a massive and expensive campaign to restore wetlands, which have the capability of recycling nitrogen back to the atmosphere.

The dead zone is not the only energy-related threat to Gulf of Mexico ecosystems. The BP/Deepwater Horizon oil spill of 2010 played out a dramatic and devastating worst-case scenario of an environmental disaster related to petroleum exploration and production. The story flourished in U.S. news coverage for several months until the well was capped. Real-time video of oil gushing into the deep sea was available for the first time ever. The gigantic oil slick that reached the ocean surface was oddly red-colored, emblematic perhaps of the wounded Earth. The well was drilled about fifty miles off the mouth of the Mississippi River, and this proximity invites a direct comparison of the potential environmental impact of biofuels versus oil production.

Estimates of the size of the oil spill vary up to about 5 million barrels, or 0.7 million metric tons. This amount, while clearly enormous, would fulfill U.S. demand for only about six hours. It is also roughly equivalent to the mass of nitrate emitted by the Mississippi and Atachafalaya Rivers during an average year. The oil spill, however, was a one-time event, although its long-term implications are still being investigated. Fertilizer-induced hypoxia has reliably recurred annually for several decades, and its long-term effects are also uncertain. Although the effects of fertilizer discharge inherently differ from those of crude oil spills, the comparison highlights the nontrivial nature of both.

Conclusions

Energy from the Sun is free, but biofuels clearly are not. Although derived from renewable crops, biofuels require substantial investment of additional, nonrenewable resources. Ironically, much of this investment presently takes the form of fossil fuels, used extensively in the production of nitrogen fertilizer, mechanized farming, and the processing of ethanol. Even if these inputs can be reduced or replaced by other alternatives, biofuel production will still require substantial input of phosphorus and potassium fertilizers. These fertilizers represent finite geological resources, although known supplies may be adequate for the next several centuries. Biofuel production also carries significant environmental costs, including soil erosion, emission of geologically stored soil carbon, and pollution of downstream water bodies by fertilizer runoff.

Soil is perhaps our most critical nonrenewable resource because, barring a revolutionary shift to aquiculture, fertile farmland is an absolute requirement for the continued survival of humans. Soils that required thousands of years to develop naturally have been converted to cropland at a geologically blinding pace. This pace cannot be sustained for long. Naturally fertile soils are restricted in geographic distribution and are genetically related to specific geologic and climatic histories. Virtually all of the most productive soils have already been enlisted for agriculture.

Future cropland expansion will largely be limited to two less attractive alternatives: drier lands that presently lie fallow or serve as pasture, and wetter lands that presently host tropical forests. The first choice can only succeed if paired with adequate water supplies, but water is itself a threatened and not entirely renewable resource (discussed in more detail in Chapter 12). The second choice may seem more viable, but will require substantial modification and maintenance of naturally infertile tropical soils. Conversion of rain forests to farmland also has potentially severe environmental consequences, in the form of decreased biodiversity and increased CO_2 emissions due to land use change.

For More Information

Burney, J. A., Davis, S. J., and Lobell, D. B., 2010, Greenhouse gas mitigation by agricultural intensification: *Proceedings of the National Academies of Science*, v. 107, p. 12052–12057.

Cordell, D., Drangert, J.-O., and White, S., 2009, The story of phosphorus: Global food security and food for thought: *Global Environmental Change*, v. 19, p. 292–305.

Dias, R. J., and Rosenberg, R., 2008, Spread dead zones and consequences for marine ecosystems: *Science*, v. 321, p. 926–929.

Donner, S. D., and Kucharik, C. J., 2008, Corn-based ethanol production compromises goal of reducing nitrogen export by the Mississippi River: *Proceedings of the National Academy of Sciences*, v. 105, p. 4514–4518.

Eglin, T., Ciais, P., Piao, S. L., Barre, P., Bellassen, V., Cadule, P., Chenu, C., Gasser, T., Koven, C., and Reichstein, M., and Smith, P., 2010, Historical and future perspectives of global soil carbon response to climate and land-use changes: *Tellus*, v. 62B, p. 700–718.

Fargione, J., Hill, J., Tilman, D., Polasky, S., and Hawthorne, P., 2008, Land clearing and the biofuel carbon debt: *Science*, v. 319, p. 1235–1238.

Foley, J. A., DeFries, R., Asner G. P., Barford, C., Bonan, G., Carpenter, S. R., Chapin, F. S., Coe, M. T., Daily, G. C., Gibbs, H. K., Helkowski, J. H., Holloway, T., Howard, E. A., Kucharik, C. J., Monfreda, C., Patz, J. A., Prentice, I. C., Ramankutty, N., and Snyder, P. K., 2005, Global consequences of land use: *Science*, v. 309, p. 570–574.

Glenn, C. R., Föllmi, K. B., Riggs, S. R., Baturin, G. N., Grimm, K. A., Trappe, J., Abed, A. M., Galli-Oliver, C., Garrison, R. E., Ilyin, A. V., Jehl, C., Rohrlich, V., Sadaqah, R. M. Y., Schidowski, M., Sheldon, R. E., and Seigmund, H., 1994, Phosphorus and phosphorites: Sedimentology and environments of formation: *Eclogae Geologicae Helvetiae*, v. 87, p. 747–788.

Goolsby, D. A., 2000(a), Mississippi Basin nitrogen flux believed to cause Gulf hypoxia: *EOS, American Geophysical Union, Transactions*, v. 81, no. 29, p. 321–327.

Goolsby, D. A., 2000(b), Nitrogen in the Mississippi Basin – estimating sources and predicting flux to the Gulf of Mexico: *U.S. Geological Survey Fact Sheet* 135-00, 6p.

Houghton, R., 2008, Carbon flux to the atmosphere from land-use changes: 1850–2005, In TRENDS: a Compendium of Data on Global Change, Technical Report, Carbon Dioxide Information Analysis Center: Oak Ridge National Laboratory, U.S. Department of Energy, Oak Ridge, TN, (http://cdiac.ornl.gov/trends/landuse/houghton/houghton.html).

Kim, S., and Dale, B. E., 2008, Effects of nitrogen fertilizer application on green-house gas emissions and economics of corn production: *Environmental Science and Technology*, v. 42, p. 6028–6033.

Lowenstein, T. K., and Spencer, R. J., 1990, Syndepositional origin of potash evaporites: petrographic and fluid inclusion evidence: *American Journal of Science*, v. 290, p. 1–42.

Montgomery, D. R., 2005, *Dirt: The Erosion of Civilizations*: University of California Press, 296 p.

Montgomery, D. R., 2007, Soil erosion and agricultural sustainability: *Proceedings of the National Academy of Sciences*, v. 104, p. 13268–13272.

Notholt, A. J. G., Sheldon, R. P., and Davison, D. F., 2005, *Phosphate Deposits of the World*: Volume 2, Phosphate Rock Resources: Cambridge University Press, 600 p.

Rabalais, N. N., Turner, R. E., and Wiseman, W. J., 2002, Gulf of Mexico hypoxia, aka "The Dead Zone": *Annual Review of Ecology and Systematics*, v. 33, p. 235–263.

Ramankutty, N., and Foley, J. A., 1999, Estimating historical changes in global land cover: Croplands from 1700 to 1992: *Global Biogeochemical Cycles*, v. 13, p. 997–1027.

Searchinger, T., Heimlich, R. P., Houghton, R. A., Dong, F., Elobeid, A., Fabiosa, J., Tokgoz, S., Hayes, D., and Yu, T.-H., 2008, Use of U.S. croplands for biofuels increases greenhouse gases through emission from land-use change: *Science*, v. 319, p. 1238–1240.

Trimble, S. W., and Crosson, P., 2000, U.S. soil erosion rates – myth and reality: *Science*, v. 289, p. 248–250.

Turner, R. E., and Rabalais, N. N., 1991, Changes in Mississippi River quality this century: *Bioscience*, v. 41, p. 140–147.

Turner, R. E., and Rabalais, N. N., 1994, Coastal eutrophication near the Mississippi River delta: *Nature*, v. 368, p, 619–621.

U.S. Geological Survey, 2010, Mineral commodity summaries 2010: *U.S. Geological Survey*, 193 p.

Wilkinson, B. H., and McElroy, B. J., 2007, The impact of humans on continental erosion and sedimentation: *Geological Society of America Bulletin*; v. 119, p. 140–156.

Zimov, S. A., Schuur, E. A. G., and Chapin, F. S., 2006, Permafrost and the global carbon budget: *Science*, v. 312, p. 1612–1613.

7

The Light of an Ancient Sun: Fossil Fuel Origins

The fuel [Egyptian railroaders] use for the locomotive is composed of mummies three thousand years old, purchased by the ton or by the graveyard for that purpose, and sometimes one hears the profane engineer call out pettishly, "D–n these plebeians, they don't burn worth a cent – pass out a King!"

Mark Twain, The Innocents Abroad (1869)

Where do fossil fuels really come from? Misconceptions about the true origin of these substances abound. One chain of gas stations uses an image of a green dinosaur as its logo; does this mean that the gasoline it sells comes from dead dinosaurs? If so, where are those rotting carcasses actually located? Directly under the gas station? The idea that oil is produced from dinosaurs is not quite so comical as it might first appear, since historically we have indeed obtained oil from the carcasses of large animals. The most famous of these are whales, some species of which were hunted nearly to extinction in the 19th century for lamp oil and lubricants. Oil in the form of animal fat continues to be rendered in large quantities for food use, but has not been widely burned for illumination since the advent of fossil fuels.

Most people's direct experience of petroleum begins and ends with the self-service pump at a gas station. The fill-up procedure is deceptively easy and quick, but in fact the hose transfers energy to your car at an astounding rate. Power is defined as the rate at which energy is transferred; delivering the energy content of 15 gallons of gasoline in 2 minutes works out to be the equivalent of about 20,000 horsepower! If you could afford to keep this flow rate going continuously it would be enough to power one engine of a medium-size jetliner. For the most part though consumers do not have direct contact with fossil energy sources at all. Natural gas, for example, travels invisibly in underground pipelines, and coal-fired power plants are often located far from the end users of the electricity they generate.

Put simply, coal, oil, and natural gas represent solar energy that was originally trapped by plants and stored in the Earth's crust. Fossil fuels can be roughly subdivided according to whether they originated from land plants, such as trees, herbs, and grasses, or aquatic plants, primarily one-celled algae. Coal deposits originate principally as land plants, and aquatic plants that lived in oceans and lakes are chiefly responsible for crude oil. As will be seen later, natural gas is a switch hitter that can be derived from any type of buried organic matter.

What about animals as a source of fossil fuel? Bones and footprints of dinosaurs are commonly found in association with coal deposits, but there is no indication that their preserved carcasses ever generated a measurable amount of oil. The potential for animals to generate petroleum is slightly greater in aquatic environments, where zooplankton, very small swimming animals, can be very abundant. Even so, aquatic plants remain by far the dominant source of oil.

The Recipe for Coal

There is nothing especially mysterious about the geologic processes involved in making fossil fuels; in fact they can be thought of in terms similar to baking a cake. You start with the right mix of ingredients, stir them together, and bake at a certain temperature for a certain amount of time (Figure 7.1). The mixing bowl for making coal is a low-lying area of standing water called a mire. Swamps and marshes are both examples of mires; trees and other woody plants distinguish swamps, whereas grasses dominate marshes. Well-known modern examples of coal-forming environments in the United States include the Atchafalaya swamp in Louisiana, the Okeefenokee swamp in Georgia, and the Everglades marsh in Florida. Areas of open water within mires also favor the growth of algae, the remains of which can make up a significant proportion of some coals.

The dead plant matter in mires is partially decayed by oxygen-breathing (aerobic) bacteria, in a process that is similar to what happens when making compost for your garden. If allowed to continue unabated, aerobic decay would completely consume all of the plant material and return its carbon to the atmosphere. However, submersion of dead plant material beneath water and burial by additional plant matter limit the available oxygen supply, and thereby slow

Coal Recipe

Start with one large swamp
(a large marsh may be substituted)

Add dead trees, bushes, and grass

Season with algae as desired

Set aside to decay until fetid

Cover liberally with thick layers of mud and sand

Bake at 100°C for 100 million years or until bituminous

Figure 7.1. The recipe for coal.

bacterial decay. Mud and sand deposited above the layer of decaying plant material seal the deal, by effectively removing the dead organics from direct contact with atmospheric oxygen.

Baking occurs by progressive addition of layers of sediment (Figure 7.2), which push the natural compost formed in mires to deeper levels of the Earth's crust where temperatures are higher. The deepest burial occurs in sedimentary basins, which serve as the oven for coal and other fossil fuels. As noted in Chapter 2, sedimentary basins are areas where layers of sand, mud, and other sediment have accumulated to great thickness. Sedimentary basins commonly occur at the edges of continents, where rivers enter the sea and discharge sediment, for example, in the Gulf of Mexico. They are also found within continental interiors, where the Earth's crust has subsided. For example, nearly the entire state of Michigan is underlain by a bowl-shaped area of sedimentary rock that reaches up to 5 km thick at its center. The deepest sedimentary basins can reach downward more than 16 km below sea level and therefore are nearly twice as deep as Mt. Everest is tall (Figure 7.3).

As a rule of thumb, temperature increases at an average rate of about 25°C to 30°C per kilometer of burial depth. In the world's deepest mine, the Rand gold mine in South Africa, natural temperatures reach a saunalike 70°C at a depth of 3,585 m. In general,

Figure 7.2. Interbedded coal and sandstone, Junggar basin, western China. Coal was deposited in swamps, sandstone deposited by rivers (photo: M.S. Hendrix).

temperatures at depths of 4 km or more exceed the boiling point of water at the Earth's surface. The water there does not actually boil, however, because of the immense pressure of overlying water and rock. The burial pressure at 4 km is about five hundred times the air

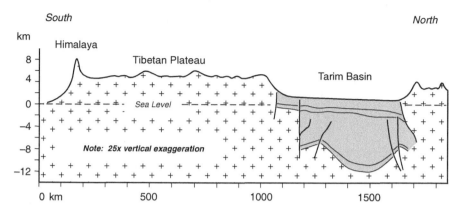

Figure 7.3. Vertical slice (cross section) from the Himalaya Mountains to Central Asia.

Gray indicates the Tarim sedimentary basin (see Figure 5.3 for location). Note that the basin extends nearly 12 km below sea level, whereas the Himalaya rise only about 8 km above sea level.

pressure used to inflate automobile tires, and five times the bursting point of typical compressed gas cylinders. The Earth might therefore be thought of as the ultimate pressure cooker. These high pressures also compress the deep plant material in coal seams to one-tenth or less of its original thickness.

The temperatures required for making coal lie well within the range of the average household oven, but if you want to cook your own coal you'll need a timer scaled in millions of years rather than hours and minutes. Coal deposits span a wide range of geologic ages, from millions to hundreds of millions of years. As it cooks, coal undergoes a series of chemical transformations, converting the original plant material into a complex natural polymer called kerogen. The chemical structure of kerogen resembles that of plastics, consisting of thousands of atoms linked together in chains and three-dimensional networks. As it heats up the coal is transformed, through a series of increasing coal ranks that have progressively greater heating values. The lowest rank is lignite, followed by subbituminous and bituminous, and culminating with anthracite (Figure 7.4). As rank increases the coal also tends to become shinier, suggesting that diamonds might form with deeper burial. However, although diamonds are in fact made of carbon, they originate at much deeper levels in the Earth and are unrelated to coal.

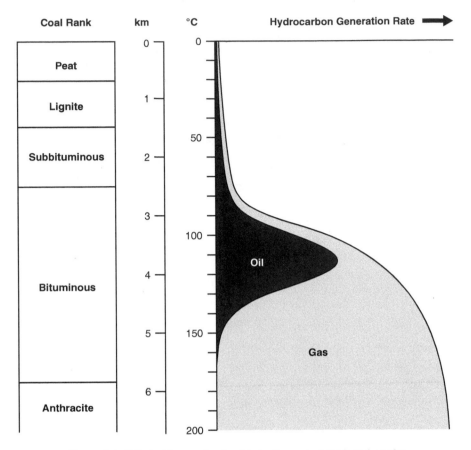

Figure 7.4. Effect of increasing burial depth on coal rank and on the generation of oil and gas from a typical marine source rock, assuming a geothermal gradient of 30°C/km.

Note that the relationships shown here are approximate and vary depending on organic matter type and heating history.

Oceans, Algae, and Oil

The recipe for oil is similar to that of coal, but with different starting ingredients. Rather than land plants growing in mires, oil starts out as one-celled algae that float near the surface of lakes and oceans. These floating algae, also called phytoplankton (Figure 7.5), absorb sunlight and store it as chemical energy. Phytoplankton therefore need to live in well-lit parts of the water column, which practically speaking means no more than about 30 m below the surface. Phytoplankton are relatively simple organisms in comparison to land plants, and because

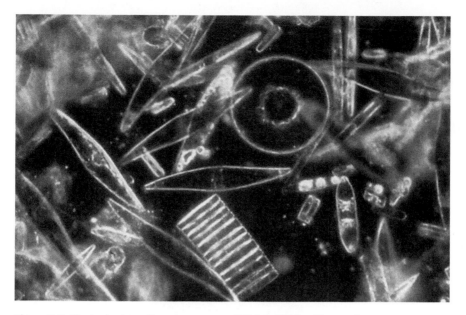

Figure 7.5. Phytoplankton (image courtesy of Richard Kirby, Plymouth University, Plymouth, United Kingdom).

they float freely they do not need to build woody support structures. They do use hydrocarbon compounds to build cell wall membranes, which regulate the flow of water, gas, and other substances between seawater (or lake water) and the interior of their cells. Phytoplankton therefore contain a relatively high proportion of hydrogen-rich organic compounds, called lipids (from the Greek word *lipos*, which means fat). These plant fats are very similar in structure to some of the molecules found in oil.

In addition to sunlight, phytoplankton need nutrients such as nitrogen, phosphorus, potassium, and others (see Chapter 6). Most of these derive from weathering of rocks on the continents and are carried by rivers (or wind) into the sea, where they can be used by phytoplankton. When phytoplankton die, they sink toward deeper waters of the ocean, taking the nutrients with them. The highest primary productivity of phytoplankton occurs where nutrient-rich waters return (or upwell) back toward the surface, commonly near the edges of the continents (Figure 7.6). Much of the open ocean is relatively dead by comparison to the continental margins. Local upwelling of nutrients from deeper parts of the ocean can cause exceptions to this pattern, for example, where the Labrador and Gulf Stream currents mix at the

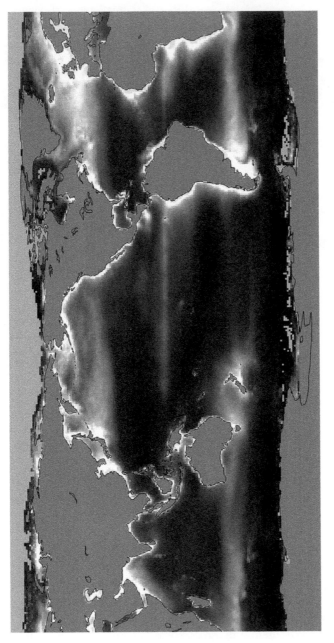

Figure 7.6. The marine "solar collector."

Lighter shades correspond to higher rates of organic productivity in the ocean, based on chlorophyll concentration. (Data collected by the Sea-viewing Wide Field-of-view Sensor on the OrbView-2 satellite from 1997 to 2002; note that no data displayed for the continents or for the Arctic Ocean [modified from image by Robert Simmon and Watson Gregg, NASA Goddard Space Flight Center Earth Observatory]).

Grand Banks offshore of Newfoundland. If the flux of nutrients to the surface of the ocean (or a lake) is large enough, it can lead to eutrophication, similar to that which occurs in the Gulf of Mexico in response to river-borne nitrogen fertilizer (see discussion in Chapter 6).

What happens next depends on where eutrophication occurs. In the open ocean oxygen may be depleted to depths as great as 1,000 m. The average depth of the ocean is about 4,000 m, however, and its deeper waters are generally well oxygenated. This is because the deep waters originate at the surface of the ocean in contact with the atmosphere. They then descend to the bottom as a result of cooling of the ocean surface at high latitudes. Eventually the deep waters rise again and complete the cycle, as part of a global conveyor belt that continuously ventilates the deep ocean (Figure 7.7). Dead organic matter that sinks below 1,000 m is mostly consumed by aerobic bacteria before it can reach the ocean floor. Sediments deposited in the really deep parts of the ocean therefore make very poor petroleum source rocks.

If eutrophication occurs near the edge of a continent, however, the resulting hypoxia may extend all the way to the seafloor. In this case dead organic matter is prevented from fully decaying. Once it has settled to the sea bottom, a continual rain of mud buries it and effectively seals it away from any additional degradation, much as plastic wrap protects food from spoiling. During periods of higher than average sea level, shallow seas can also spread across low-lying areas of continental interiors. These shallow seaways can become depleted in oxygen, resulting in deposition of widespread source rocks for oil and gas (Figure 7.8).

Abundant organic matter can also be preserved in lakes, where there is no global conveyor belt to replenish the water with oxygen. Anyone who has dived to the bottom of a midwestern lake in the summer has likely experienced firsthand the murky, organic-rich ooze that lines the bottom (along with assorted "fossils" including beer cans, boat parts, and the occasional whole snowmobile). In fact, seasonal overturn of some euthrophic lakes can even kill fish that live close to the surface, by mixing of oxygen-depleted water with shallower parts of the lake.

Once buried in mud, phytoplankton and other organic remains are cooked in a manner similar to coal. Unlike coal, however, aquatic organic matter initially generates liquid oil rather than natural gas. You might expect crude oil to look something like the used motor oil from the crankcase of a car, and sometimes it does. However, crude

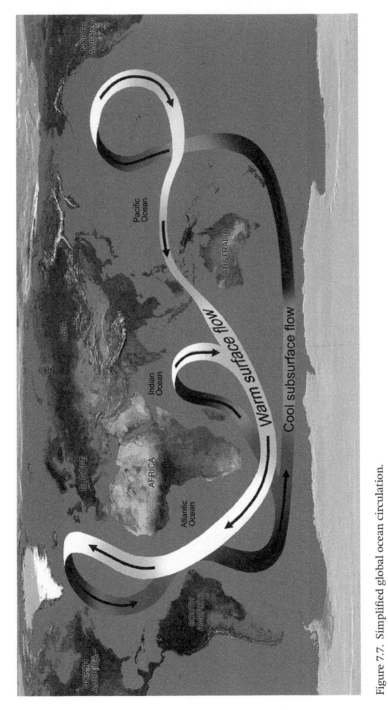

Figure 7.7. Simplified global ocean circulation.

Deep circulation of cold waters helps to keep the ocean floor well oxygenated, preventing burial of organic matter anywhere other than continental margins (NASA/JPL).

Figure 7.8. Organic-rich mudstone containing about 10 percent organic matter by weight, lying above a lighter colored rock layer.

oils span a surprisingly wide range of compositions, to the extent that it is difficult to formulate a single precise definition of this common substance. Naturally occurring oils range in color from pitch-black to completely transparent, with various shades of brown, yellow, and even red being common. Oil can occur as a waxy solid at surface conditions or as a watery liquid. Some oils are so dense they sink in water, whereas others may be as light as gasoline. Some oils have little or no smell, but others are so sulfurous that the slightest whiff will quickly clear a room.

Some crude oil actually starts out as a gas at reservoir conditions, but then condenses to a clear liquid at the surface. Called condensate, this liquid has a chemical composition not unlike that of refined gasoline. Oil field workers have been known to pilfer such oil, colloquially known as "drip gas," for direct use in their trucks. There is no one standard crude oil, although certain oils such as West Texas Intermediate or North Sea Brent have been designated as benchmarks for price comparison. The actual price per barrel of any particular crude oil may differ from these benchmarks, depending on its characteristics. The highest prices are paid for relatively light, sweet (low sulfur) crudes, whereas heavy or sour (sulfur-rich) crudes bring lower prices.

Crude oil can in general be defined as a naturally occurring solution, in which the solute consists of relatively large molecules composed of hydrogen and carbon, as well as nitrogen, sulfur, oxygen, and other trace elements. The solvent consists of relatively small hydrocarbon molecules. This solution is staggeringly complex, containing tens of thousands of distinct chemical compounds. Crude oil as it emerges from the ground needs to be refined. This is done primarily by separating different components of this complex mixture on the basis of their boiling points. The underlying principle is simple: Light oil fractions with relatively small molecules (fewer carbon atoms) boil at lower temperatures, whereas heavy oil fractions with relatively large molecules (many carbon atoms) boil at higher temperatures. To start the process, crude oil is heated at the base of a reactor tower, causing some of the oil to vaporize. As it rises through the tower it begins to cool and condense, at which point it is extracted. Products with low boiling points, such as gasoline, rise to the top of the tower before condensing. Products with high boiling points, such as heating oil and asphalt, remain near its base.

Natural gas may also be dissolved in oil, but most of this gas is separated before the oil is pumped into the reactor. Some small amount of gas typically remains, however, and for safety reasons is flared at the top of the tower. Additional refinery steps are used to convert some of the heavier fractions chemically into lighter, more valuable products, particularly gasoline. This is done by chemically breaking down large molecules into smaller ones, a process called cracking, and by hydrotreating to add hydrogen.

The Many Faces of Natural Gas

Natural gas is a marketing department's dream: it is all natural, organic, clean burning (except CO_2 emissions), and clear. In its natural state it is also odor free; the familiar gas smell results from an artificial additive. Without the additive natural gas would be imperceptible to humans, and gas leaks could easily result in asphyxiation or explosions. Methane, which consists of one carbon and four hydrogen atoms, is the most common component of natural gas. Gas molecules that contain two, three, or four carbon atoms (ethane, propane, and butane, respectively) constitute most of the remainder.

Natural gas might be thought of as the small change of the fossil fuels world, which is returned from a variety of transactions that occur during burial of organic matter. The first of these is the transformation of dead plant material into coal. As coal is heated it becomes more and more carbon-rich, through the loss of its other main constituents oxygen and hydrogen. These changes are accomplished by release of water, carbon dioxide, and methane. The cumulative release of methane during coalification can be quite large, equivalent to up to 10–20 percent of the total energy content of a bituminous coal. At greater depths and pressures, phytoplankton-rich source rocks can also generate substantial amounts of methane, typically after they've already generated most of their oil (Figure 7.4). Finally, even oil itself can generate gas, if it experiences elevated reservoir temperatures for long periods, by thermal cracking of large molecules into smaller ones.

Small quantities of natural gas seem to occur in wells drilled just about anywhere, even in places where other fossil fuels are absent. Methane can also be produced biologically by microorganisms that live in swamps, landfills, and even the stomachs of cows. However, gas produced at elevated temperatures in the Earth's crust, called thermogenic gas, accounts for virtually all commercial production.

Non-hydrocarbon gases are also part of the mix, and most notably include carbon dioxide and nitrogen. Sour gas contains hydrogen sulfide (H_2S), a deadly substance that has a rotten egg odor at low concentrations. At higher concentrations it can overwhelm the human sense of smell and become difficult or impossible to detect. Hydrogen sulfide kills invisibly, taking its victims quickly and by surprise. On the lighter side, helium also occurs in natural gas; its best known physiological effect is to make you talk like Donald Duck. Helium originates primarily by radioactive decay of uranium and thorium deep in the Earth. Concentrated natural helium accumulations are rare, and in fact most of the current U.S. supply of helium is produced from natural gas wells in the Texas Panhandle.

From the Bottom Up?

Virtually all geologists agree that fossil fuels contain energy from the Sun that has been buried beneath the surface of the Earth. A small minority of scientists believe the opposite, however: that natural gas and perhaps even crude oil originate at great depth, without the need of living precursors. The idea of an abiogenic (nonbiological) origin for petroleum has been around in one form or another since at least the 19th century. It seems to attract increased attention whenever oil prices are high, since generation of oil and gas deep within the Earth might indicate that the available supplies are considerably larger than usually assumed.

For most crude oil the abiogenic origin theory is easily exposed as inadequate, because many of the molecules in oil have intricate chemical structures that can be linked directly to similar compounds found in organisms living at the Earth's surface. For example, consider the family of compounds called steroids, best known for their use by some athletes to help build muscle mass. All steroids contain the same basic chemical structure, consisting of three rings with six carbon atoms each, and one ring with five carbon atoms. Steroids occur naturally in the human body (for example, in cholesterol) and in plants. They are also found in most crude oil, in forms that can be related to their plant precursors. Steroids and other biomarker compounds cannot plausibly have arisen from a deep, abiogenic source.

It is also commonly observed in oil fields that progressively fewer large hydrocarbon molecules are preserved with increasing depth, and that at the deepest levels only natural gas and light oils remain. This suggests that crude oil generally does not survive the temperatures typical of the deeper parts of the Earth's crust, greater than approximately 175°C–200°C. The final stake through the heart of the abiogenic oil hypothesis may be the simple fact that oil fields have only been found within or in proximity to sedimentary basins. Rocks that formed at temperatures greater than 200°C generally do not contain any oil, except where they have been brought into contact with sedimentary rocks that do.

There is more room for negotiation with respect to natural gas. Laboratory experiments have shown that methane and other similarly simple hydrocarbon compounds can be synthesized at high temperatures and pressures, through various inorganic mineral reactions. Methane and ethane have also been detected in places where a biological origin is extremely unlikely, such as comets and on Titan, the largest moon of Saturn. However, the important resource question is not whether abiogenic methane exists, but whether commercial quantities of hydrocarbon fuels on Earth can be attributed to such an origin. Most geologic evidence and the collective learning of nearly 150 years of commercial oil and gas exploration say this is not likely.

The most famous advocate for a deep-seated origin of petroleum was not a geologist but an astrophysicist, the late Thomas Gold. Gold proposed that oil and gas could be formed by microbes capable of living at elevated temperatures and that world oil supplies might therefore be much larger than generally believed. Fittingly, these ideas were tested at the intersection between astronomy and geology. A well was drilled in 1985–1986 at an ancient meteorite impact site in Sweden called Siljan Ring. Its proponents hoped that the impact would have caused deep cracks in the Earth's crust, and that these cracks allowed oil and gas to migrate closer to the surface. This site is in an area underlain by granite, and thus unlikely to have any oil sourced from organic matter contained in sedimentary rocks.

Ultimately the findings from the well proved as controversial as its underlying hypothesis. Small amounts of oil were found but were immediately suspected of representing contamination from the drilling fluid that had been pumped into the well. Gold claimed that larger amounts of naturally occurring oil had been produced but never presented evidence supporting this claim. No additional wells have been drilled to test the nonbiologic hypothesis since Siljan, and Sweden still relies exclusively on imports to satisfy its petroleum needs.

Efficiency of the Geologic Solar Collector: Part I

The efficiency of any energy source can be simply defined as the percentage of the total available energy that can be used to generate electricity, propel an automobile, or satisfy other practical needs. For example, the efficiency of a photovoltaic solar collector can be defined as the electrical energy produced in a given time, as a percentage of the total solar energy striking its surface. The rated efficiency of photovoltaic cells is typically rather low – in the range of 10–20 percent conversion of sunlight into electricity – and their practical efficiency may be lower still. Most of the incident sunlight becomes wasted heat.

Following this same logic we can calculate the efficiency of the Earth itself as a giant solar collector. Although it doesn't directly produce electricity, the Earth actively stores energy in its crust in the form of fossil fuels. Most of this storage occurred during the Phanerozoic eon, which represents the most recent 541 million years of Earth history (see Figure 2.5). The efficiency of the Earth as a solar energy collector can be calculated by comparing the total energy stored in fossil fuels to the amount of sunlight that reaches Earth in 541 million years (see Table 7.1). The result of this calculation is rather startling: The efficiency of the Earth as a solar collector is on the order of only 4 billionths of 1 percent! Put differently, the energy in fossil fuels that

Table 7.1. *The Earth as a solar collector*

Sunlight Reaching Earth		
Sunlight intensity at Earth's surface[1]	1×10^3	W/m^2
Cross-sectional area of Earth	1.3×10^{14}	m^2
Total solar power reaching Earth's surface	1.3×10^{17}	W
Solar energy reaching Earth per day	1.1×10^{22}	J
Solar Energy Storage in Fossil Fuels		
Energy content of known fossil fuels[2]	9×10^{22}	J
Efficiency of solar energy storage	0.000000004	%
Fossil fuel - sunlight equivalent	*~1*	*week*
Total Solar Energy Storage in Earth's Crust		
Energy content of all buried organic carbon[3]	6×10^{26}	J
Efficiency of solar energy storage	0.00002	%
Total carbon storage - sunlight equivalent	*~100*	*years*

Notes:
[1] Average solar intensity with the Sun directly overhead on a cloudless day.
[2] Based on known, recoverable fossil fuels estimated by Rogner and others (2013).
[3] Based on Berner (2001, 2003) isotopic mass balance model of organic carbon burial rates over past 541 million years. Organic matter assumed to contain 75 percent carbon by weight, and to have an energy value of 17.5 MJ/kg.

accumulated over the past 541 million years is roughly the equivalent of 1 week of sunlight striking the Earth.

Why is the Earth such a laughably poor solar collector? The explanation starts with the fact that plants are not particularly good at collecting sunlight, giving both biofuels and fossil fuels a built-in disadvantage. Matters go downhill quickly from there. The vast majority of dead plant matter never gets a decent burial and instead is left to rot on the surface. Of the plant matter that is buried and becomes incorporated into rocks, most of it is too scattered and dilute to generate concentrated oil accumulations. In many cases where organic matter *has* been buried in sufficient concentration to generate petroleum, it is buried either too deeply or not deeply enough. It might also be prematurely exhumed during the uplift of mountains and exposed to erosion.

Furthermore, much of the oil and gas that is actually generated does not remain trapped in rocks. Instead, it leaks all the way to the surface as natural seeps. Oil seeps are a common feature of the coastal waters of southern California, for example, and seeps led to the original historical discovery of petroleum as a fuel. Once at the surface, coal, oil, and natural gas cannot survive long in the presence of free oxygen. They are quickly destroyed, returning their carbon to the atmosphere as CO_2.

Given all these obstacles one might wonder how oil, gas, and coal have become so abundant. A clue exists in the famous multiple monkey theorem, which postulates that one thousand monkeys typing on one thousand typewriters for a long enough time will eventually reproduce the complete works of William Shakespeare. The details and attribution of this theorem vary, but at its heart lies the idea that even the most improbable events can occur if given enough time. The Earth may be an exceptionally poor solar collector, but it has been operating for an exceptionally long time.

Efficiency of the Geologic Solar Collector: Part II

The preceding calculation depends on accurate knowledge of the amount of recoverable fossil fuels, but history has shown this number to be a moving target. Estimates of the world's present endowment of fossil fuels vary widely and are subject to considerable controversy. Fossil fuels that are not presently profitable to extract may become so in the future as a result of higher energy prices or the development of new extraction technologies (see further discussion in Chapter 15). Resources such as deeply buried coal seams and offshore methane hydrate accumulations are known to be immense in magnitude, but their potential to be profitably extracted is largely speculative.

It is therefore impossible to draw a definite line between presently economical fossil fuel reserves and resources that might become economical in the future. However, an alternative way to calculate the efficiency of the geologic solar collector is to take into account all of the energy stored by organic matter in the Earth's crust, regardless of whether or not it can eventually be extracted. This calculation establishes the upper theoretical limit for how high the geologic solar efficiency can be.

There are two basic approaches to the problem. One is to do a global-scale accounting of all sedimentary rock formations that contain organic carbon and multiply their volume by the percentage of organic carbon they contain. Another approach involves the relative abundance of two different isotopes of carbon, which contain twelve versus thirteen neutrons in their nuclei. Plants prefer to incorporate ^{12}C into their tissues rather than ^{13}C, with the result that preserved organic matter has a systematically lower $^{13}C/^{12}C$ ratio than does inorganic CO_2 in the atmosphere or ocean. Because the total quantity of each isotope doesn't change much, increased organic matter burial causes a corresponding shift in the $^{13}C/^{12}C$ ratio in the ocean. This ratio can be measured via the isotopic composition of carbon in marine limestone ($CaCO_3$), and used to estimate the rate at which organic carbon was buried at various times in geologic history (see Figure 2.7).

Both of these approaches give similar answers. The total energy in dead plant matter buried in the Earth's crust is the equivalent of about 100 years of sunlight striking the Earth (see Table 7.1), which is *5,200 times* greater than the energy contained in known fossil fuels! This difference is hugely important to our understanding of the ultimate potential of fossil fuel resources. Unlike the various estimates that have been made of economic fossil fuel abundance, the total amount of buried organic carbon represents a real physical limit of the Earth. Very little of the total amount of organic matter in the Earth's crust will ever be profitable to extract, but its enormous magnitude proves that we have barely begun to scratch the surface of naturally stored solar energy. The ultimate limits of economic fossil fuel production will therefore depend not only on how much has already been proven to exist, but also on how far we can afford to dig into the Earth's much larger warehouse of "fossil sunlight."

Conclusions

Fossil fuels are really fossil biofuels; they contain solar energy that was first captured by plants and then buried in the Earth's crust. As an energy collection system, fossil fuels combine the relatively low efficiency of plants as solar collectors with the far worse efficiency of the Earth in storing dead plant matter. This dreadfully low efficiency is offset by the availability of vast amounts of geologic time, however, resulting in the accumulation of substantial fossil fuel resources. The dramatic changes in human lifestyle that began with the Industrial Revolution were made possible by exploiting these resources at rates vastly greater than their formation. Unfortunately, the combustion of fossil fuels also liberates carbon dioxide at rates far greater than those at which it was originally trapped by plants and buried in the crust.

The total amount of plant matter buried in the Earth's crust is clearly finite, raising the question of how much longer we can continue to support ourselves using these nonrenewable resources. This question will be examined in more detail in Chapter 15. For now, it is worth reflecting on the fact that the total amount of fossil sunlight stored in the Earth's crust is thousands of times greater than the perceived magnitude of economic fossil fuels. The former quantity is fixed, but the latter quantity has grown continuously for most of the past century and a half.

For More Information

Berner, R. A., 2001, Modeling atmospheric O_2 over Phanerozoic time: *Geochimica et Cosmochimica Acta*, v. 65, p. 685–694.

Berner, R. A., 2003, The long-term carbon cycle, fossil fuels and atmospheric composition: *Nature*, v. 426, p. 323–326.

Glasby, G. P., 2006, Abiogenic origin of hydrocarbons: An historical overview: *Resource Geology*, v. 56, p. 85–98.

Gold,T., and Soter, S., 1980, The deep earth gas hypothesis: *Scientific American*, v. 242, p. 155–161.

Gold, T., 1992, The deep, hot biosphere: *Proceedings of the National Academy of Science*, v. 89, p. 6045–6049.

Hubbert, M. K., 1956, *Nuclear energy and the fossil fuels*: Houston, Texas, Shell Development Company Publication #95, 40 p.

Klemme, H. D., and Ulmishek, G., 1992, Effective petroleum source rocks of the world: Stratigraphic distribution and controlling depositional factors: *American Association of Petroleum Geologists Bulletin*, v. 75, p. 1809–1851.

Kroeger, K. F., Primio, R., and Horsfield, B., 2011, Atmospheric methane from organic carbon mobilization in sedimentary basins – the sleeping giant?: *Earth-Science Reviews*, v. 107, p. 423–442.

Rogner, H. H., Aguilera, R. F., Archer, C. L., Bertani, R., Bhattacharya, S. C., Dusseault, M. B., Gagnon, L., Haberl, H., Hoogwijk, M., Johnson, A., Rogner, M. L., Wagner, H., Yakushev, V., Arent, D. J., Bryden, I., Krausmann, F., Odell, P., Schillings, C., and Shafiei, A., 2013, Energy resources and potentials, *in GEA, 2012: Global Energy Assessment – toward a Sustainable Future*: Cambridge University Press, Cambridge, UK and New York, NY, and the International Institute for Applied Systems Analysis, Laxenburg, Austria., p. 425–512.

Royer, D. L., 2006, CO_2-forced climate thresholds during the Phanerozoic: *Geochimica et Cosmochima Acta*, v. 70, p. 5665–5675.

Schoell, M., 1988, Multiple sources of methane in the Earth: *Chemical Geology*, v. 71, p. 1–10.

Tissot, B. P., and Welte, D., 1984, *Petroleum Formation and Occurrence*: Heidelberg, Springer Verlag, 699 p.

8

Digging for Daylight: Coal and Oil Shale

FO'SSIL. *adj.* [*fossilis*, Latin; *fossile*, French] That which is dug out of the earth.

COAL. *n.f.* [col, Sax.; *kol* Germ. *kole*, Dutch. *kul*, Danish] The common fossil fewel.

Excerpts from Dictionary of the English Language,
Samuel Johnson, 1755

The word "fossil" is taken from the Latin verb *fodere*, which means to dig. In modern usage "fossil" refers to any remnant, imprint, or trace of past life. Some common examples include dinosaur bones, buried clamshells, or mineralized wood. The most abundant fossils also turn out to be the smallest: shells and other remains of one-celled organisms that lived in the surface waters of the ocean and in lakes. More broadly, the term "fossil" also includes things like footprints, burrows, or leaf impressions that have been preserved in rocks.

On the basis of the modern definition of fossil, it is debatable whether the term "fossil fuel" accurately describes coal, petroleum, and natural gas. There is virtually no doubt that these substances originated in living organisms, but it is generally difficult to associate them with any unique biologic precursors. Instead they represent the highly modified remains of various different organisms that have been commingled into a kind of generic stew. Rather than being analogous to, say, a dinosaur skeleton, fossil fuels are more like a dinosaur that has been run through a meat grinder, bones, teeth, and all, along with several other types of dinosaurs and whatever smaller animals happened to wander by. It would be a clever curator who could reconstruct a meaningful museum exhibit from the resulting fossil sausage.

The term "fossil" was originally used much more broadly, however, to mean *anything* that was dug out of the Earth. This included not only fossils in the modern sense, but also metals, ore minerals, rock salt, gemstones, and certain materials that can support combustion. Coal, the most notable example of the latter, has been used as a heat source for millennia. Oil shale is somewhat less well known, but its use as a fuel has similarly ancient beginnings. It is clear from Sam Johnson's dictionary that the term "fossil fuel" was already in use by the middle of the 18th century and possibly earlier, and in 1755 the term "fossil" did not necessarily imply biologic origin. Coal, oil shale, crude oil, and natural gas therefore all qualified as flammable fossils. Paleontologist Martin Rudwick wrote in 1976 that "it was not until the early nineteenth century that the word 'fossil', without qualification, finally became restricted to [the biological] end of the spectrum."

The first fossil fuels to be exploited were the easiest to find and dig up and could be used interchangeably with firewood. Coal commonly occurs in widespread deposits near the Earth's surface. In some cases the coal may be exposed at the surface, either naturally or in excavations made for railroad lines or roads. In other cases it leaves a telltale sign in the form of clinker, a distinctive brick-red discoloration of rock layers that results from the natural burning of coal. Most deep mines are developed in areas where coal was initially discovered at shallow depths. The general location of significant coal deposits is rarely in question. This knowledge often dates back centuries; for example, the phrase "carry coals to Newcastle" has been used to indicate a pointless or futile activity since at least the 17th century, at which time the existence of major coal deposits in northern England was already ancient knowledge.

Major areas of coal production are sometimes referred to as coalfields, a term that evokes fields of wheat, corn, or other agricultural crops. Such analogies are in fact rather appropriate, in part because coal originated as plants and in part because coal beds commonly spread out across large areas. Coal must also be laboriously harvested, via surface or underground mining.

By definition, oil shale is any rock from which oil may be economically extracted by heating in the absence of oxygen, a process known as retorting. Retorting effectively simulates natural oil generation from source rocks, but at higher temperatures (typically 350°C –500°C) and over shorter periods. Oil shale can also be directly combusted, but its relatively high mineral content normally makes this impractical. Like coal, oil shale is abundant near the Earth's surface and its potential has been known for centuries. The requirement for heating it adds substantial cost, however, and has thus far inhibited large-scale commercial development.

The Making of Coal

Although easy to find, coal is not ubiquitous. It requires specific geologic conditions for its formation. The key to coal is the long-term persistence of mires, as discussed in the previous chapter. This means that the land surface and the water table must remain in close spatial association with each other over long periods, in order to sustain swampy or marshy conditions. In addition, continuous slow subsidence is required to protect thick coal seams from erosion. Two natural settings are especially conducive to coal formation: coastal river deltas and closed drainages within the continents. In both cases an abundant supply of freshwater is required to promote vigorous plant growth and to maintain a high water table. Coal-forming environments therefore are often associated with relatively humid climates.

Coal-forming coastal deltas occur where major rivers meet the sea. As discussed in Chapter 4, rivers can be thought of as machines whose primary job is to carry sediment to the ocean. Their ability to move sediment is related directly to their velocity: Swifter currents can move small rocks and even boulders, whereas slow-moving rivers carry only mud and sand. When a river finally meets the ocean its job is done. Its velocity quickly decreases, and its sediment load is unceremoniously dumped. The area where the dumping occurs is called a delta, in honor of the triangular coastal plain built by the Nile River (the Greek historian Herodotus noted that this plain resembles an inverted letter delta (Δ)). Not all deltas share the same shape, but all lie at elevations very close to sea (or lake) level. Deltas are therefore ideal factories for the production of coal.

The vast coal reserves in the United Kingdom, Germany, and the eastern United States that sparked the Industrial Revolution originated on soggy river deltas around 300 million years ago. Several factors conspired to create coal at geologically unprecedented rates. A northern continent that included present-day North America collided with a southern continent that included Africa, Europe, and parts of Asia, raising an enormous mountain range that ran roughly

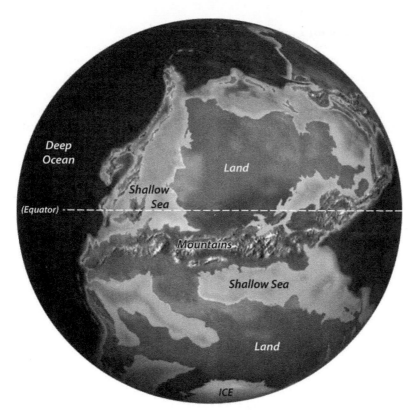

Figure 8.1. Configuration of the supercontinent Pangea at approximately
300 million years ago.

Note east-west oriented mountain chain near the equator and glacial ice cap
surrounding the South Pole (Ron Blakey, Colorado Plateau Geosystems).

east-west along the equator (Figure 8.1). Remnants of this mountain
range are preserved today in the Appalachian Mountains of the eastern United States and the Atlas Mountains of Morocco. The weight
of the mountains pushed down the Earth's crust in immediately
adjacent areas, causing them to subside and be flooded by seawater. Rivers carrying sediment from the mountains drained into these
shallow seas, depositing large coastal deltas. High rates of equatorial precipitation helped ensure humid conditions and promoted
the growth of lush vegetation. Continued burial of this plant matter
eventually converted it into bituminous and anthracite coals with
relatively high heating values.

Sweet and Sour Coal

The adjectives "sweet" and "sour," as applied to crude oil, refer to variations in the amount of sulfur it contains. Sweet crudes contain low amounts of sulfur, making them easier to refine and generally more valuable. Sour crudes, on the other hand, reek of rotten eggs because of their higher sulfur content and cost more to refine. Coals likewise vary widely in sulfur content. Unfortunately the abundant Carboniferous coals of the eastern United States and other similar deposits must on average be considered sour, because they contain relatively high amounts of sulfur. Sulfur is released during combustion as sulfur dioxide (SO_2), which can then combine with water to form sulfuric acid (H_2SO_4), the principal culprit in acid rain.

The sulfur in Carboniferous coal is directly related to its coastal origins. Sulfur that is leached from the continents and emitted from volcanoes builds up over time in the ocean, largely in the form of the sulfate ion (SO_4^{2-}). Coastal mires commonly contain a brackish mixture of fresh river water and seawater, and therefore the peat they form is bathed in marine sulfate. Mere exposure to sulfate might not sour coastal coals, but sulfate opens the door for its unsavory associates, the sulfate-reducing bacteria. These characters are among oldest known organisms on Earth, perhaps appearing as early as 3.5 billion years ago. They owe their antiquity to the fact that the early atmosphere of the Earth contained no free oxygen; sulfate-reducing bacteria therefore evolved the ability to breathe sulfate instead. They exhale pungent hydrogen sulfide, giving them an extreme case of bad breath. There is less use for this ability now that the atmosphere contains 18 percent oxygen, but it still comes in handy from time to time. For example, aerobic bacteria quickly consume all the available oxygen in coastal peat mires. Sulfate-reducing bacteria can continue to thrive in such environments after the oxygen is gone, and the hydrogen sulfide they emit either combines with iron to form either the mineral pyrite (FeS_2, also known as "fool's gold") or sulfur-bearing organic compounds. As a result, coal that evolves from brackish mires often contains more than 3 percent sulfur by weight.

In contrast, sweet coals have managed to maintain a low-sulfate lifestyle and in so doing have largely evaded the sulfate-reducing bacteria. Whereas high-sulfur coals generally formed where rivers entered the sea, most low-sulfur coals formed where rivers entered inward-draining topographic basins located within the continents. Commonly such basins also contain lakes, which vary in character from isolated ponds scattered across low-lying floodplains to huge overflowing expanses of uninterrupted freshwater. Precipitation and river influx constantly replenish these lakes and mires, which in turn continually spill water to downstream rivers. There is therefore little time for sulfate to accumulate. As a result, coals formed in these settings often contain less than 1 percent sulfur. Low-sulfur coals are common in the western United States, and recently mining has been especially active in the Powder River Basin of Wyoming.

A Hole in the Mountains: The Powder River Basin

During the mid-19th century hundreds of thousands of settlers traveled from the central United States to the Northwest, via a series of wagon routes known collectively as the Oregon Trail. Rugged mountains in Colorado, Montana, and Idaho obstructed travel through those areas, channeling emigrants across the less intimidating high prairies of Wyoming. Wyoming has mountains too but they are less continuous, separated by broad expanses of more subdued topography. These flat areas are typically underlain by thick accumulations of sedimentary rock and thus mark the locations of major sedimentary basins.

The geologic history of these basins traces back about 70 million years to the Cretaceous period, when mountainous basin walls first began to isolate what was previously a shallow inland sea. Rivers draining the eroding mountains carried sediment to build up flat basin floors and maintained water tables close to the land surface. The resultant mires deposited truly spectacular quantities of peat, which with continued burial matured into enormously thick beds of low-sulfur, subbituminous coal.

Travelers of the Oregon Trail followed the North Platte River westward from Scott's Bluff, Nebraska, into what is now Wyoming, skirting the southern edge of a wide grassy region known as the Powder River Basin (Figure 8.2). As they trudged along the rutted trail they could not have appreciated the abundance of coal beneath the high plains to the north, which were then tribal lands of the Sioux, Cheyenne, and Arapahoe. Mining of Powder River Basin coal began in the late 19th century. For many decades its production remained minuscule, for two good reasons. First, its heating value is only about two-thirds that of higher-rank bituminous and anthracite coals. Second, coal in Wyoming or Montana shares something in common with solar power in Arizona or wind power in North Dakota: All three

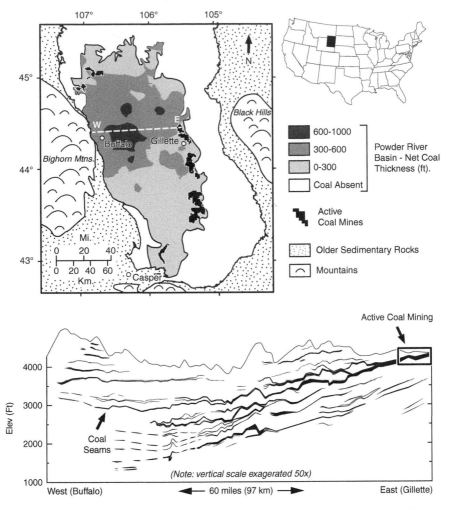

Figure 8.2. Distribution of coal in the Powder River Basin, Wyoming and Montana.

The lower part of the figure shows an east-west cross section of the coal deposits beneath the dashed white line shown on the map (modified from Flores et al., 2010).

require lengthy transmission infrastructures to carry them to distant markets. The power lines for coal consist of parallel sets of steel rails that carry railroad cars rather than moving electrons. Powder River Basin coal sold at the mine is literally dirt cheap, costing only one-fifth to one-sixth as much as some eastern coals, but the train ride to the east can easily double or triple its cost.

Political and economic developments that began in the 1970s brought about a reversal of fortune for Powder River Basin coal. Passage of the U.S. Clean Air Act of 1970 and its later amendments required reduced emissions of sulfur and other pollutants from coal-fired electrical plants. One way to achieve reductions in sulfur emissions is to install scrubbers that remove SO_2 from flue gases, through a chemical reaction with limestone ($CaCO_3$). This reaction converts gaseous SO_2 into solid calcium sulfite ($CaSO_3$). Calcium sulfite can be converted into calcium sulfate (gypsum), the principal ingredient in wallboard. Scrubbers can be very effective at removing sulfur before it enters the atmosphere, but they substantially increase power generation costs. They also release additional CO_2 derived from the limestone.

Coal from the Powder River Basin contains much less sulfur and therefore costs less to burn, giving it a distinct competitive advantage. Railroad deregulation in the late 1970s further encouraged the use of Wyoming coal, through long-term reductions in the costs of transporting it. Production skyrocketed between 1970 and 2008, while production of eastern coals on average remained unchanged. Currently the Powder River Basin by itself produces about 40 percent of the coal burned in the United States, raising questions about its sustainability. At first glance such concerns appear unfounded because the total in-place coal deposits are immense. Recent assessments by the U.S. Geological Survey conclude that there are about 1 trillion tons of coal in-place, which if readily available would satisfy U.S. consumption requirements for at least several centuries.

However, the same U.S. Geological Survey studies concluded that only about 25 billion tons of coal can be economically extracted using current mining techniques. This is less than 1.5 percent of the total resource and represents only another twenty years or so of production at current rates. Depending on your point of view this conclusion could be perceived as either witheringly pessimistic, from the perspective of resource exhaustion, or blithely optimistic, from the perspective of CO_2 emissions. But is it accurate?

Hiking to Buffalo

To better understand the difference between in-place coal and commercially extractable resources you might take a leisurely stroll between the towns of Gillette and Buffalo, Wyoming, a distance of about sixty miles as the crow flies (Figure 8.2). At the beginning of this walk you will encounter huge open-pit mines, railroad lines, and the people and infrastructure required to support both. A single hundred-foot-thick coal seam named the Wyodak Coal supports most of the mining near Gillette. Its thickness varies with location, and in some mines it has split into two separate seams named Anderson and Canyon. Variable amounts of sandstone and mudstone, generically referred to as "overburden," lie above the coal and must be stripped away; the ratio of overburden to coal is referred to as the "stripping ratio." The overburden thickness around Gillette can reach up to two hundred feet, giving an average stripping ratio of around 2:1. The excavated overburden is used to backfill older mine cuts, and then sculpted into an imitation of the original topography. To date, about 6 billion tons of coal has been mined.

As you walk westward from Gillette the overburden gradually thickens and the mines give way to semiarid scrubland, which is pockmarked by thousands of wells drilled to recover methane from coal lying beneath the surface. The U.S. Geological Survey has arbitrarily defined the western boundary of the Gillette coalfields as a north-south line that passes about twelve miles west of town. At this point the depth to the coal interval that is being mined near Gillette has increased to something close to one thousand feet. The geology has also become considerably more complicated; the Wyodak Coal has split into several thinner seams, and a number of new coal seams have now appeared above and below this interval. In 2008 the U.S. Geological Survey mapped eleven major named coals that have been penetrated by wells and estimated a total of 201 billion tons of coal in place in near Gillette. Assuming a maximum stripping ratio of 10:1 and other practical considerations, the U.S. Geological Survey calculated that about 77 billion tons of coal could potentially be recovered by surface mining. This amount of coal would satisfy total U.S. demand at current rates for more than half a century.

As you stand amidst the sagebrush pondering these numbers you may begin to reflect on the enormity of the mining operations you witnessed near Gillette and then begin to wonder about the costs of removing hundreds of feet of *additional* overburden to get at the coal underneath. After all, an estimate of technically recoverable coal is meaningless unless the coal can be sold for more than it cost to mine.

The U.S. Geological Survey has thought of this as well. In January 2007, when they were compiling data for their report, Powder River Basin coal sold for an average of about $10.50 per short ton. At this price they estimated that only about 10 billion tons of Gillette-area coal remains to be economically mined, suggesting that more than one-third of the original economic reserves in this area have already been depleted! This shocking realization must be tempered, however, with the knowledge that market prices change continuously, and therefore so do reserves. By March of 2008 the price for Powder River Basin coal had increased to $14.00 per short ton; leading the U.S. Geological Survey to estimate that economic reserves had nearly doubled to 18.5 billion tons. At prices greater than $60 per ton, similar to the contemporary price of Appalachian coal, the U.S. Geological Survey estimated that the full 77 billion tons of technically recoverable coal might be economical to mine!

The remaining fifty miles of your walk to Buffalo crosses some truly lonely and monotonous territory, populated mostly by wildlife, cattle, sheep, and gas wells. Your attention is more likely to be drawn by the towering Bighorn Mountains looming ahead, which mark the western boundary of the Powder River Basin. As you look up, the largest part of the Powder River coal deposits are passing beneath your feet. Extending their earlier trend, coal seams continue to divide and multiply going westward. The aggregate thickness of these individual seams increases at least sixfold, compared to the single major seam being mined in Gillette. The U.S. Geological Survey has estimated that areas outside the Gillette coalfields hold 654 billion tons of coal in place. Most of this coal lies beyond the reach of surface mining, however, and some of it would be impractical to extract even with underground mining. How then should the practical magnitude of coal reserves be assessed? There really is no single answer to this question, because it depends strongly on future prices and on future extraction technology. Probably none of this coal is economically recoverable at current prices. However, the U.S. Geological Survey estimated that underground mining at depths of one thousand to two thousand feet might become profitable at prices in the mid-$30s per ton. Currently there are no underground mines in the basin, but their future development cannot be categorically ruled out.

By the time the sleepy town of Buffalo pulls into view your brain may feel slightly addled from trying to grasp the enormity and complexity of coal in the Powder River Basin. Or, perhaps you're just feeling heatstroke from walking three days in the Wyoming wind and sun. Either way, you will have gained an appreciation that fixed opinions about the size of coal reserves are overly simplistic at best. The real question for the future is not whether coal supplies will be exhausted, which is virtually impossible, but how much coal remains to be mined at prices we can afford to pay.

The Consequences of Coal

Coal offers two distinct advantages as a fuel: It has a high energy density, and it is relatively cheap to obtain. Unfortunately the downside of coal can be equally dramatic, starting from the point at which it is extracted from the ground. Although its safety record has been steadily improving with time, underground mining of coal has long been considered the exemplar of a dirty and dangerous job. Surface mining inevitably causes major environmental disturbance because the preexisting environment must, by definition, be removed in order to get to the coal beneath (see Figure 8.3). Of the various types of surface mining, "mountaintop removal" in West Virginia and adjacent areas has become the most infamous. It permanently alters the preexisting topography, by excavating the highs and infilling the lows. Surface mining on flatter terrain is more amenable to reclamation of the preexisting land surface by backfilling the original excavation. Reclamation cannot truly restore the landscape to its original state, however, because natural land surfaces evolve through complex interactions of geology, climate, and living organisms over periods of thousands of years or more.

Some of the more insidious environmental damage caused by both surface and underground coal mines stems from the fact that coal naturally concentrates both sulfur and toxic metals such as arsenic, cadmium, lead, mercury, and many others. In fact, a useful analogy can be drawn to a home water filter that uses activated charcoal, to remove impurities contained in the local drinking water supply. Undisturbed coal in the ground does much the same thing, retaining chemical impurities in a relatively immobile state. However, mining of coal tends to release toxic substances, in part because the coal is exposed to oxygen. Oxygen, aided by iron-oxidizing bacteria, causes pyrite and other sulfide minerals to decompose, releasing sulfur. Sulfur in turn reacts with water to form sulfuric acid, which is the principal cause of acid mine drainage. These waters also carry heavy metals released by oxidation.

Burning of coal releases whatever toxic substances remain after mining. As noted previously, much of the liberated sulfur can be captured before it leaves the smokestack. Other pollutants, in particular heavy metals, may be carried away as part of the "fly ash" that is more difficult to contain. As a result, coal burning is currently the largest source of environmental mercury and arsenic pollution. More surprisingly, it is also the largest source of radioactivity released to the environment

Figure 8.3. Open pit coal mine near Point of Rocks in southwestern Wyoming (note trucks on Interstate Highway 80 in the foreground for scale).

by humans, because coal naturally sequesters uranium and thorium. A well-known 1978 study by J. P. McBride and coauthors from Oak Ridge National Laboratory concluded that people living near coal-fired power plants typically receive higher radiation doses than those living near nuclear power plants (although the doses were very low in both cases). More ominously, the combustion of coal emits up to twice as much CO_2 per unit of generated energy as combustion of other fossil fuels, making it a leading contributor of atmospheric greenhouse gases.

Spirits of Coal: Synthetic Gas and Oil

In the early 19th century the city of London began to illuminate its streets with gas derived from the heating of coal. Synthetic gas, or "town gas," as it was sometimes called, consists of a flammable mixture of hydrogen, carbon monoxide, CO_2, and water vapor. Because it could be efficiently distributed through a network of pipes it represented a revolutionary advance on previous street lighting systems. Town gas was also used for many years as a fuel for stoves and ovens. The efficiency of using coal-derived gas for lighting or heating is questionable, however, because considerable energy must be expended to heat the coal to high temperatures. However, coal was plentiful and cheap.

In the 1920s the German chemists Franz Fischer and Hans Tropsch took synthetic gas a step further, by inventing a process to produce ersatz (imitation) liquid fuels. Hydrogen and carbon monoxide were starting ingredients, and cobalt or other metals were used as catalysts. Their invention was especially important in a country with abundant coal but little petroleum. By the latter years of World War II the Nazi army had become increasingly desperate for transportation fuel, the shortage of which contributed to their ultimate downfall. By this time an estimated 9 percent of their total petroleum usage was supplied by the Fischer-Tropsch process. Unfortunately the generation of liquid fuels from coal is also thermodynamically disadvantageous, because large amounts of heat must be expended to convert synthetic gas to hydrocarbon liquids. It has also been estimated that the use of synthetic fuels derived from coal approximately doubles emissions of CO_2 compared to those of conventional petroleum. Since World War II, commercial production of synthetic liquids from coal has been limited to South Africa, where Apartheid era trade embargoes restricted availability of petroleum.

Coal gasification has recently experienced renewed interest as a way to improve the overall efficiency of coal-fired power plants; the process is based on the combustion of synthetic gas at very high temperatures in gas turbines and secondary use of waste heat to generate steam. Underground coal gasification has long been considered as a way of extracting energy from coal without the need to mine it, essentially by setting the coal on fire and collecting the gases that result. The environmental effects of such a process would no doubt be challenging to control, however, and CO_2 emissions would likely be large.

Fire and Water: Coal Bed Methane

Coal miners learned early on to fear underground explosions, caused by an invisible gas known as firedamp. The Davy lamp, invented in 1815, was designed to reduce this risk by using an enclosed flame that was less likely to ignite flammable gases. It also burned more brightly when such gases were present, providing an early warning to miners. Firedamp consists mostly of methane, which may originate either from thermal degradation of coal as it is heated during burial or from microbial degradation of coal at lower temperatures.

Methane tends to adsorb to the surface of coal particles, held there by relatively weak electrostatic forces. Note that *adsorption* refers to the adhesion of molecules on a surface and is different from *absorption*, which occurs when liquid or gas is retained inside a solid (for example, a sponge). Lint sticking to a sweater is a good analogy for adsorption; the lint is the equivalent of methane and the sweater is the coal. Typically only a single layer of methane molecules can be adsorbed onto the surface of an organic particle in coal. However, the total amount of methane that can be adsorbed by coal is very large, because the total surface area of organic particles is large and because the methane molecules pack themselves closely together.

The key to packing this much methane onto coal surfaces is pressure; the higher the pressure, the more methane can be adsorbed, up to a point. What matters to methane is not the pressure on the coal itself, but water pressure within its pores. Even in their compacted form the coals still contain many small holes and fractures, which are filled with water. The weight of water in pore spaces lying above the coal exerts pressure, in proportion to the thickness of the overlying water-saturated interval. The situation is exactly analogous to a swimming pool, where pressure increases continuously as you dive deeper, except that the pore spaces in coal are far too small to swim through. Water pressure in the pore spaces in coal serves to keep methane molecules adsorbed onto organic particles.

So long as the water stays in the coal the methane does too. However, if water is drained to allow underground mining, the pore pressure within the coal drops and methane can "desorb" and escape into mine tunnels. The earliest coal bed methane wells were drilled in an effort to lessen this hazard, by removing some of the methane before mining commenced. The methane so produced could also be sold to recoup the costs of drilling. By the 1980s coal beds were being drilled solely for their methane reserves, rather than for improvement of mine safety. Coal bed methane also carries some significant downsides, the greatest of which is the need to pump large volumes of water out of the coal before significant gas production can occur. The water in some coal seams circulates relatively rapidly from the surface and is essentially fresh, making it safe (probably) to release directly into surface streams. More saline and potentially toxic waters tend to occur in deeper coals; these waters must either be treated or else reinjected into deeper, saline aquifers. Either solution adds to the overall cost of obtaining methane.

Other problems stem from the fact that gas doesn't flow very easily through coal. In fact the economic viability of coal bed methane depends strongly on the presence of numerous small fractures in the coal, called cleats, that provide avenues for gas to travel to a wellbore. Coal bed methane development also commonly requires drilling a lot of wells. Nearly thirty thousand wells have been drilled to date in the Powder River Basin, spaced as close as 400 m apart. Such dense drilling adds more cost and results in substantial land surface disruption both from the wells themselves and from the many roads, pipelines, and facilities that accompany them.

Oil Shale: The Fuel of Last Retort?

Imagine that you are on a camping trip and have built a fire circle out of local rocks. You've cooked your dinner and have let the fire burn down, but before you drift off to sleep you make a curious discovery: Even though the fire in the pit has been reduced to embers, some of the rocks surrounding it still appear to be burning! At some point in early human history this scene likely played out in reality, leading to the accidental discovery of oil shale. Later on, more enterprising sorts learned that not only can certain rocks sustain combustion, but that when baked at temperatures in the range of 400°C they give off noxious vapors that condense into a dark sticky oil. This experiment is easy to reproduce in any high school chemistry lab, by heating some crushed oil shale in the bottom of a test tube over a Bunsen burner. The resultant vapors will condense on the cooler upper end of the test tube. Unfortunately the resultant smell, which is reminiscent of burning tires, may also empty the classroom.

In simplest terms, oil shale is a rich petroleum source rock that has not yet generated its petroleum. As discussed in greater detail in Chapter 10, shale is just mud, which accumulated on the bottom of lakes or the ocean and has been gradually compressed by the weight of overlying rocks. Shale is particularly good at preserving the dead remains of microscopic organisms, which gradually transform into kerogen. Oil generation can occur naturally if kerogen-rich shale is buried to a depth of a few kilometers, typically at temperatures in the range of 80°C–140°C (see Figure 7.4). This requires some patience, however, as it may require millions of years for the requisite amount of burial and heating to occur.

It is possible to have your oil now, however, if you're willing to work for it. Oil shale has been mined from the ground for centuries and then heated in retorts (a kind of oven) to generate oil. Retorting differs from combustion in that it doesn't require oxygen; in fact, oxygen must be deliberately excluded from the process to the extent possible. Heating without oxygen breaks chemical bonds in the kerogen, causing it to fall apart into smaller organic molecules that are similar to those contained in naturally generated oil.

The origins of this oil shale retorting are difficult to trace; however, a British patent issued in 1694 (or 1684 according to some sources) describes "a way to extract and make great quantities of pitch, tarr, and oyle out of a sort of stone." It is unclear whether the inventors were able to capitalize on their patent, but by the mid-19th century oil shale deposits in France, Germany, and Scotland were being commercially retorted and refined, with the oil used primarily as an illuminant. An active oil shale industry was established in the Lothian region of Scotland by 1850 and persisted for more than a century. Oil shale was mined at the surface and underground at depths as great as 365 m and heated in a variety of retort designs that used coal as the primary heat source. In its early days of exploitation the Scottish oil shale yielded as much as 40 gallons of oil per ton of rock, but yields had dwindled to half of this by the time mining ended in 1962.

Today very little oil is produced from oil shale; the leading producer is Estonia, a tiny Baltic country that produced about 400,000 barrels of oil in 2011 (equal to about 0.01 percent of overall global oil production). The resource potential of oil shale is enormous, however. Dr. John R. Dyni of the U.S. Geological Survey estimated in 2006 that oil shale deposits worldwide hold about 2.8 *trillion* barrels of oil. This is roughly equal to the U.S. Geological Survey's estimate of the world's original endowment of conventional oil, and more than triple the historical total world oil production to date. About half of this total was from the Green River Formation in the western United States.

Eocene Great Lakes: Oil Shale in the Western United States

The landscape of northwestern Colorado, southwestern Wyoming, and eastern Utah evokes every Western movie ever made. Distant snow-capped peaks provide the backdrop for emerald river bottoms, dry sagebrush-dotted plains, high mesas, and imposing cliff faces, with an occasional patch of sandy desert thrown in for dramatic effect. Fifty million years ago conditions were quite different, however. The area

Figure 8.4. Fossils preserved in Green River Formation lake deposits
Upper left: *Cockerellites liops*, approximately 13 cm long. Upper right: softshell
turtle 1.7 m long. Lower left: 2 m long *Palmites* sp., indicative of a climate
similar to modern Florida. Lower right: *Procambarus primaevus*, a crayfish
approximately 5 cm long that lived in shallow near shore waters (U.S.
National Park Service).

was divided into broad basins like the Powder River Basin, but instead
of coal swamps they were occupied by a series of large lakes comparable in size to the Great Lakes that currently border the American
Midwest and Canada. The climate at that time was more equable than
it is today, and a remarkable diversity of plants and animals lived on
the lakeshores, including palm trees, crocodiles, and the diminutive
ancestors of modern horses. The water in the Eocene Great Lakes var-
ied from being fresh and teaming with fish and other aquatic animals
(Figure 8.4) to being several times saltier than the ocean and devoid
of any but the most brine-tolerant of organisms. The latter included
specialized bacteria and flamingo-like ducks that dined on them.
Freshwater lakes upstream commonly drained to more saline lakes
downstream, and the entire assemblage of lakes slowly changed as
regional climate and river courses evolved (Figure 8.5).

The Eocene Great Lakes expanded and contracted many times during their history, depositing thick beds of mud rich in calcite ($CaCO_3$),

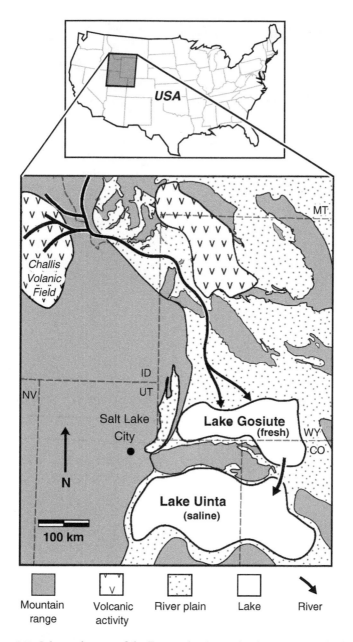

Figure 8.5. Schematic map of the Eocene landscape in the western United States approximately 49 million years ago.

A major river system that originated on a high-elevation volcanic plateau in central Idaho drained southeast into freshwater Lake Gosiute in southwestern Wyoming, which in turn spilled south into saline Lake Uinta (based on Smith et al., 2008).

Figure 8.6. Eocene Green River Formation outcrop in the Piceance Creek sedimentary basin of northwestern Colorado.

Exposure has approximately 650 m of relief. Darker beds at the bottom are reddish-colored river plain deposits; lighter color beds with distinct horizontal banding are oil shale of varying richness. Top surface is capped by resistant sandstone, derived from volcanic rocks in central Idaho.

dolomite $(CaMg)(CO_3)_2$, and kerogen. Green River Formation oil shale deposits often reach 20 percent or more organic matter by weight and hundreds of meters in preserved thickness (Figure 8.6). The lakes also experienced intermittent periods of intense evaporation, during which they deposited prodigious amounts of sodium carbonate and bicarbonate salts (including the minerals trona and nahcolite), which are mined for use in glassmaking, baking soda, and other applications.

Recently revisions by R. C. Johnson and colleagues at the U.S. Geological Survey suggest that the total in-place oil resources of the Green River Formation in Utah, Colorado, and Wyoming may together reach 4.3 trillion barrels (Dyni's 2006 estimate only considered the deposits in Colorado). However, there is a catch. These estimates are based solely on what is in the ground, not on what can be practically recovered. Estimates of the recoverable amount of oil are largely speculative, since commercial production has not yet been established. Michael D. Vanden Berg of the Utah Geological Survey pointed out in 2008 that only a small fraction of the in-place amount is ever likely to be recoverable, however. Assuming that economic oil shale beds need to be at least 5 feet thick, lie at burial depths of less than 3,000 feet, and yield at least 25 gallons of oil per ton of rock, Vanden Berg concluded that about 77 billion barrels of oil are potentially recoverable from Green River Formation oil shale in Utah. This is only about 6 percent of the U.S. Geological Survey's 2010 estimate of 1.32 trillion barrels of oil in the same area. Applied to the Green River Formation over its entire tristate area, a 6 percent recovery factor would still imply total recoverable reserves of more than 250 billion barrels, however, roughly equivalent to the reported oil reserves of Saudi Arabia.

Early commercial attempts to exploit Green River Formation oil shale were hampered by the costs incurred by surface mining and retort operations. More recent approaches avoid mining altogether and instead focus on retorting oil shale in its natural underground setting. Such "in situ" techniques hold two major technical advantages over surface retort operations. First, rocks provide better insulation than any surface retort container can ever hope to achieve. The process heat needed for suretort therefore can be more effectively conserved. Second, slow underground heating of oil shale over a period of years produces a higher-quality product, similar in its initial composition to kerosene or jet fuel. The less desirable tar and other by-products associated with surface retorting are left in place.

A remarkable diversity of in situ heating schemes have been devised, and several of these have been expanded to the stage of field experiments. Such schemes must overcome several challenging hurdles to heat the oil shale in a way that permits oil to be recovered profitably, while avoiding contamination of freshwater supplies. The best known approach is the Shell In situ Conversion Process (ICP), which uses electrical resistance heaters placed in vertical wells to cook the oil shale gradually at temperatures around 350°C over a period of two to four years. The generated oil is then pumped out of separate producing wells. The heating and producing wells would both be isolated from local groundwater systems by a "freeze wall," created by using an outer ring of wells spaced 8 feet apart. Chemical refrigerant circulated through the outer wells would freeze nearby groundwater, creating an underground curtain of ice. This scenario might be thought of as the opposite of Baked Alaska, a famous dessert created by quickly baking meringue around a core of ice cream. In Shell's scheme the center is heated and the ice is on the outside.

Shell is not alone in exploring oil shale extraction; for example, ExxonMobil and the American Shale Oil Company have invested in alternative in situ technologies. While various approaches have shown promise, none is yet near the point of commercial oil production. Years of further research and development still lie ahead, with the pace of this activity dependent on future oil prices.

Oil Shale Energy Return and CO_2 Emission

For all its potential bounty, oil shale will always be handicapped relative to naturally liquid oil deposits because the latter have in effect been subsidized by geothermal heat, which has already transformed kerogen into oil. In contrast, the extraction of oil from oil shale will always require a substantial up-front investment of energy. How well does oil shale pay off this investment? The full answer to this question is complex and depends in part on whether combustion of natural gas obtained from retorting of the oil shale itself is counted as one of the energy inputs. Dr. Adam Brandt, a researcher at Stanford University, has produced some of the most careful peer-reviewed studies of this topic to date, based on analyses of both surface and in situ retorting processes. Disregarding input energy obtained from the oil shale itself, he calculated "external energy ratios" (EER) ranging between 2.4 and 15.8 for the Shell ICP process. These numbers roughly agree with informal industry estimates that the "energy returned on investment" (EROI; see Chapter 15 for further discussion) for oil shale is something like 3:1. These numbers fall short of the generally accepted range of EROI for most conventional oil fields but are higher than the energy return of corn ethanol.

Beyond questions of its net energy return, another major attraction of oil shale is its potential to convert solid fuels such as coal, which is used primarily to generate electricity, into liquid transportation fuel. The U.S. Navy acknowledged the importance of oil shale as a potential liquid fuel source as early as 1909, when it designated deposits in the western United States as a strategic petroleum reserve to help ensure a sustainable supply of fuel for its ships. U.S. coal deposits were already known to be large, but many major oil field discoveries still lay in the future.

Unfortunately oil shale development also bears some of the same environmental burdens as coal, such as the potential for surface disruption and contamination of freshwater supplies. It is also expected that oil shale development will consume large quantities of water, principally used for remediation of spent shale to remove pollutants. Estimates for the actual amounts of water needed vary widely. Perhaps oil shale's most daunting issue, though, is CO_2 emissions, because the process heat needed for retorting will most likely be from combustion of fossil fuels. Oil shale of the western United States and some other countries also contains large amounts of the minerals calcite and dolomite, which can decompose into CO_2 during retorting. Taking all of these factors into account, Brandt calculated that the overall CO_2 emissions associated with in situ retorting would be 21 percent to 47 percent higher than those from conventional petroleum-based fuels. He estimated that surface retorting is an even worse option, resulting in a 50–75 percent CO_2 penalty over conventionally produced gasoline.

Conclusions

Coal and oil shale represent fossil fuels in their simplest form: the raw remains of plants that have been buried and preserved in rocks. As such they are also the most ubiquitous of fossil fuels; coal in particular has been an inexpensive staple of human energy use for centuries. Barring major economic or political changes, it seems likely to continue in this role for a long time to come.

Oil shale has also been around for centuries, but because it contains substantially greater amounts of noncombustible mineral matter it has never been as useful as coal for direct combustion. It holds a big advantage over coal, however, in its ability to generate liquid hydrocarbons directly. Even if it were not a net energy source on its own, oil shale might still serve as an intermediary for converting energy obtained from sources such as nuclear fission or even solar energy into liquid form.

Finding coal and oil shale deposits is a relatively trivial task compared to exploring for conventional oil and gas fields. The locations of most major coal and oil shale deposits are already well known, and the magnitude of in-place resources is immense. The magnitude of potentially recoverable resources is not as easily defined, however, because this depends on future energy prices and extraction technology.

Finally, both coal and oil shale pose significant environmental problems, for which there seem to be no really elegant solutions. In particular they both suffer from higher CO_2 emissions than other fossil fuels.

For More Information

Allix, P., Burnham, A., Fowler, T., Herron, M., Kleinberg, R, and Symington, B., 2010, Coaxing oil from oil shale: *Oilfield Review*, v. 22, p. 4–15.

Brandt, A., 2008, Converting oil shale to liquid fuels: Energy inputs and greenhouse gas emissions of the shell in situ conversion process: *Environmental Science & Technology*, v. 42, p 7489–7495.

Cane, R. F., 1976, The origin and formation of oil shale, *in* Yen, T. F., and Chilingar, G. V., eds., *Oil Shale*: Amsterdam, Elsevier, p. 27–60.

Carroll, A. R., and Wartes, M. A., 2003, Organic carbon burial by large Permian lakes, Northwest China, *in* Chan, M. A., and Archer, A. W., eds., Extreme depositional environments: Mega end members in geologic time: *Geological Society of America Special Paper* 370, p. 91–104.

Dyni, J. R., 2006, Geology and Resources of Some World Oil Shale Deposits: *United States Geological Survey Scientific Investigations Report* 2005–5294, 49 p.

Flores, R. M., and Bader, L. R., 1999, Fort Union Coal in the Powder River Basin, Wyoming and Montana: A Synthesis: *United States Geological Survey Professional Paper* 1625-A, 75 p.

Flores, R. M., Spear, B. D., Kinney, S. A., Purchase, P. A., and Gallagher, C. M., 2010, After a Century – Revised Paleogene Coal Stratigraphy, Correlation, and Deposition, Powder River Basin, Wyoming and Montana: *United States Geological Survey Professional Paper* 1777, 106 p.

Hawkins, D. G., Lashof, D. A., and Williams, R. H., 2006, What to do about coal?: *Scientific American*, September, p. 68–75.

Jevons, W. W., 1865, *The Coal Question: An Inquiry Concerning the Progress of the Nation, and the Probable Exhaustion of Our Coal-mines*: London and Cambridge, MacMillan, 349 p.

Johnson, R. C., Mercier, T. J., and Brownfield, M. E., 2011, Assessment of In-Place Oil Shale Resources of the Green River Formation, Greater Green River Basin in Wyoming, Colorado, and Utah: *United States Geological Survey Fact Sheet* 2011–3063, 4 p.

Leckel, D., 2009, Diesel production from Fischer-Tropsch: The past, the present, and new concepts: *Energy and Fuels*, v. 23, p. 2342–2358.

Luppen, J. A., Scott, D. C., Osmonson, L. M., Haacke, J. E., and Pierce, P. E., 2013, Assessment of Coal Geology, Resources, and Reserve Base in the Powder River Basin, Wyoming and Montana: United States Geological Survey Fact Sheet 2012–3143, 6 p.

McBride, R. E. Moore, J. P. Witherspoon, and R. E. Blanco, 1978, Radiological Impact of Airborne Effluents of Coal and Nuclear Plants: *Science*, v. 202, p. 1045–1050.

Rudwick, M. J. S., 1976, The Meaning of Fossils; Episodes in the History of Paleontology (2nd ed.): New York, Science History Publications, 287 p.

Schweinfurth, S. P., 2009, An Introduction to Coal Quality: United States Geological Survey Professional Paper 1625-F, 20 p.

Smith, M. E., Carroll, A. R., and Singer, B. S., 2008, Synoptic reconstruction of a major ancient lake system: Eocene Green River Formation, Western United States: *Geological Society of America Bulletin*, v. 120, p. 54–84.

Vanden Berg, M. D., 2008, Basinwide evaluation of the uppermost Green River Formation's oil shale resource, Uinta basin, Utah and Colorado: *Utah Geological Survey Special Study 128*, 19 p.

9

Skimming the Cream: Conventional Oil and Gas

> Petroleum has of late years become the matter of a most exten-
> sive trade, and has even been proposed by American inventors
> for use in marine steam-engine boilers. It is undoubtedly supe-
> rior to coal for many purposes, and is capable of replacing it.
>
> *William Stanley Jevons, 1865*

Perhaps the most iconic image of the bounty of fossil fuels was the
well drilled at Spindletop Hill in east Texas, which began to gush oil
in the opening days of the 20th century (January 10, 1901; Figure 9.1).
Oil spewed more than 150 feet in the air, at an initial rate estimated
at approximately 100,000 barrels per day. This flow rate translates to
a power output of about 6 gigawatts, roughly equivalent to the rated
capacity of *three* Hoover Dams (power is defined as the rate at which
energy is transferred). Remarkably, the liberation of this staggering
energy resource only required drilling 1,139 feet into the Earth, a
depth less than that of many municipal water wells. Once unleashed, it
took nine days to bring the gusher under control. With such immense
power waiting literally to leap from the ground, it is small wonder
that oil went on to reshape the world.

Uncontrolled eruptions of oil like the one at Spindletop are
of course hugely wasteful and pose great danger due to their flam-
mability and to the contamination of the surrounding area. The
petroleum industry and government regulators therefore go to great
lengths to prevent such catastrophes, which have ironically become
all the more shocking for their rarity. The fundamental geologic
truth of conventional oil fields remains unchanged more than a century after Spindletop. All such accumulations comprise fluid hydrocarbons that have been highly concentrated by natural geologic

Figure 9.1. Gusher at Spindletop, Texas, in 1901 (Anonymous).

processes. These concentrated fluid accumulations lend themselves to comparatively easy (and controlled) extraction, encouraging an analogy to the cream that can be skimmed from the top of a pail of raw milk.

Locating concentrated natural oil and gas accumulations is anything but easy, though, since by definition these treasures are relatively small and lie hidden thousands of feet below the surface of the Earth (or ocean). Their discovery therefore has always relied on a variable blend of observation, inference, and luck, aided by the application of large sums of cash.

The Earth as a Sponge

Jules Verne's 1864 science fiction novel *A Journey to the Center of the Earth* spun a fanciful tale of a swiss-cheese-like planet, riddled with caverns that housed giant mushrooms, dinosaurs, and even a subterranean ocean. The protagonists entered this hidden world through a volcanic crater in Iceland and somehow exited through the active Mediterranean volcano Stromboli. The reality is considerably less exciting; large open caverns do exist but they seldom extend more than a few hundred meters below the surface. Krubera Cave in the Caucasus Mountains holds the current world record for deepest known cave at 2,191 m. Although impressive, this depth represents only about 0.03 percent of the mean distance from the surface to the center of the Earth, which is about 6,378 km.

The notion that the Earth is substantially hollow can be easily disproven on the basis of what we know of its mass and the composition of its rocks. However, the uppermost reaches of the Earth's crust are not entirely solid either; sedimentary rocks in particular are riddled with many small holes, called pores. These holes may be thought of as comparable to the holes in a sponge, and in aggregate they may represent a substantial volume. It is within these pores that oil and gas reside (Figure 9.2). The sizes of individual pores vary widely, but most fall in the range of nanometers to millimeters. The most sponge-like rocks are sandstone, limestone, and their close sedimentary relatives. Sandstone consists of broken and abraded bits of other rocks; its porosity resides in the spaces between those fragments (Figure 9.2). Limestone can likewise contain broken fragments of coral, shells, or other bits of calcium carbonate grown in the ocean, resulting in sandstonelike porosity and permeability. Limestone porosity also may be enhanced by dissolution of calcium carbonate by acidic waters, which taken to the extreme results in the formation of caves.

The proportion of holes to total rock volume (known as porosity) in a sandstone may start out as high as 40 percent or more at the surface. Porosity decreases systematically with increasing burial, because of more efficient packing of sand grains, collapse of pores under the weight of overlying rocks, and aqueous precipitation of new minerals that clog porosity. Typical oil- and gas-bearing sandstone beds at moderate burial depths of a few kilometers commonly have porosities in the range of 10–25 percent. At depths greater 10–15 km most porosity has disappeared, effectively eliminating the possibility that we can drill deeper wells in search of oil and gas.

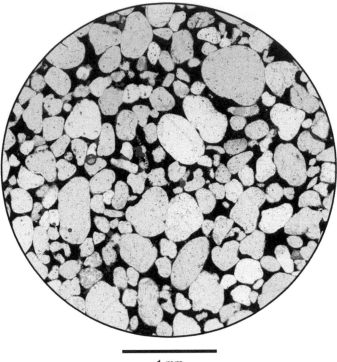

1 mm

Figure 9.2. Microscopic view of a sandstone.

Light gray objects are sand grains; black indicates pore spaces between grains that can be filled with water, oil, or gas (porosity equals approximately 27 percent of the volume of this sandstone).

Verne's subterranean ocean is perhaps not quite so fanciful as it first appears; most subsurface pores are in fact filled with water, and most of this water is at least as salty as the ocean. A relatively thin zone of freshwater (aquifers) floats above this subterranean ocean, filling the pores closest to the Earth's surface. Oil and gas are comparatively minor subsurface constituents; they are generally less dense than either saline or freshwater and therefore buoyant in the subsurface. Once generated, they therefore rise toward the surface of the Earth, completing the analogy to cream rising to the top of a pail of milk.

The Ascent of Hydrocarbons: Oil and Gas Migration

Oil and gas can only rise through rocks that are truly spongelike, meaning not only that they have pores but that the pores are connected in

Figure 9.3. Sandstone cliffs of the Paso del Sapo Formation, near the Chubut River in Argentina.

a way that allows fluid to flow between them. This quality, called permeability, allows water to enter a sponge in the first place, or to be expelled when a sponge is squeezed. Paradoxically, the initial generation of oil and gas typically occurs within organic-rich mudstone that lacks significant permeability. Hydrocarbons do manage to escape, however, and are then free to begin their upward journey.

Oil and gas generally migrate along relatively continuous beds of permeable rock such as sandstone (Figure 9.3). These beds do not necessarily provide the shortest path to the surface, but they do provide the fastest. Major faults and fractures encountered along the way may offer shortcuts through which hydrocarbons can traverse impermeable beds and shorten the trip. The length of the journey varies greatly, from a few hundred meters to a few hundred kilometers. Long-distance flow of hydrocarbons might be compared to the drainage of water through from a river valley, except that the underground rivers run uphill. Both systems serve to concentrate relatively modest flows that originated over a wide area, into larger, more focused streams.

Left to their own devices fluid hydrocarbons would continue their ascent until they reached the Earth's surface, where their prospects for long-term survival would be poor. Contact with water, oxygen, and microorganisms degrades oil relatively quickly, turning it into gooey tar or solid bitumen, a polymeric substance resembling black plastic. With long enough exposure even the solid bitumen is oxidized, releasing its carbon to the atmosphere as CO_2. Natural gas meets the same fate much more quickly, since it more readily dissipates into atmosphere. Fortunately, the same processes that continually destroy naturally seeping oil also help to mitigate the impact of unnatural spills from oil wells, pipelines, and tankers. Much of the oil that was spilled from the BP/Deepwater Horizon well in 2010, for example, is believed to have naturally biodegraded in the warm surface waters of the Gulf of Mexico.

It appears that the majority of fluid hydrocarbons generated in the Earth's crust do eventually seep to the surface. Surface oil seeps have been recognized for millennia and used for sealing, lighting, paving, and even medicinal purposes. Natural gas seeps are also common; for example, one such seep supports a well-known "eternal flame" on the Indonesian island of Java. Gas seeps are also abundant in the Middle East and Caspian Sea regions and may have helped to inspire ancient Zoroastrian religious beliefs concerning eternal fire. The total rate of seepage is poorly known, but Keith Kvenvolden of the U.S. Geological Survey and his colleagues estimated in 2003 that approximately 600,000 tons of oil per year seeps into the ocean. Natural gas seepage is even faster; Kvenvolden and colleagues estimated in 2005 that 45 million tons of gas seeps into the atmosphere each year.

These natural seepage rates are relatively small compared to current rates of commercial fossil fuel production; 600,000 tons per year represents only about 0.01 percent of commercial crude oil production, and 45 million tons represents only about 2 percent of natural gas production. However, if projected over long periods they add up to substantial totals. Oil seeping continuously at 600,000 tons per year would match the magnitude of currently proven world oil reserves within about 400,000 years. Gas seepage would equal proven gas reserves in a mere 2,500 years! The uncertainties in these estimates are large and the rates may not be constant through time, but it is clearly a very leaky system.

The earliest oil wells were shallow hand-dug pits, which served to expose less-degraded oil immediately below surface seeps. By the mid-19th century these efforts began to be mechanized, using cables suspended from wooden derricks to operate primitive well-digging tools. The first such wells were drilled in the 1840s to 1850s, in a number of localities including Azerbaijan, Poland, Romania, Canada, and the United States. These generally were not deliberate attempts to find the sort of highly concentrated hydrocarbon accumulations responsible for the blowout at Spindletop. Instead they merely facilitated existing natural flows of oil from surface seeps.

The Making of a Gusher

Despite the intrinsic leakiness of sedimentary basins, a small fraction of generated oil and gas does remain trapped below the Earth's surface. The simplest traps are those where rock layers have been folded by tectonic forces into an underground arch, called an anticline (Figure 9.4). An anticline acts as a sort of upside-down bowl that catches ascending oil and gas before it can reach the surface. A wide variety of other geologic irregularities produce the same net effect, such as lateral gradation of permeable reservoir sandstone into impermeable mudstone or juxtaposition of a sandstone bed against mudstone across a fault. Traps are what make highly concentrated accumulations of oil and gas possible.

To better understand the physics of an oil gusher, also known as a blowout, it helps to think of a waterbed. In its simplest form a

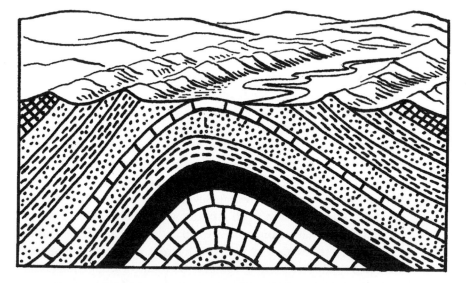

Figure 9.4. Schematic illustration of an anticline (Pearson Scott Foresman, Wikimedia Commons).

waterbed consists of a large plastic bag filled with water and supported by a rigid frame. Waterbeds reached their peak popularity during the 1970s, shortly after the occurrence of peak oil production in the United States but before the Baby Boom generation got old enough to develop bad backs. The water in a waterbed can be compared to the fluids in a porous and permeable petroleum reservoir, and the rigid bed frame is analogous to the rocks that constitute the remainder of its bulk.

Imagine that you puncture the top of an unoccupied waterbed with a knife or other sharp object. Some minor leakage will no doubt occur, but otherwise not much will happen because the water pressure at the top of the bed is about equal to atmospheric pressure. The water pressure inside the bed increases downward at a predictable rate, in proportion to the weight of overlying water. So long as there are no restrictions to flow the situation is stable, with no tendency of water to flow up or down. The same would be true in a permeable sandstone reservoir, which is filled entirely with oil all the way to the surface. A well drilled into such a reservoir cannot gush and would in fact need to be pumped.

Now imagine instead that someone is lying on top of the waterbed. The bed has to support the weight of its occupant in addition to the weight of overlying water, which it can do if the bed remains sealed. However, if you stab the bed with a knife (taking care not to injure its occupant), water will gush upward through the hole. Something very similar happens in the case of an oil well blowout. Playing the role of plastic sheeting are impermeable rocks called seals (or sometimes caprock), which typically consist of salt or mudstone (Figure 9.5). These impermeable rocks inhibit the upward flow of fluids from underlying reservoir rocks. Overlying rocks play the role of the people lying on the water bed.

Impermeable seals can cause reservoir pressures in an oil or gas field to rise above the surface exerted by the weight of overlying water (or oil) alone, a condition known as overpressure. The end result is a reservoir whose contents may violently erupt to the surface if its seal is penetrated by a careless driller. Several other factors may also contribute to the violence of such eruptions, including compressive tectonic stresses, fluid volume increases related to oil and gas generation, buoyancy forces exerted by reservoir hydrocarbons, and the regional flow of groundwater from topographic highs into a sedimentary basin.

Once a blowout has begun, natural gas dissolved in the oil helps to sustain it. If the pressure on a reservoir is relieved by the escape of

Figure 9.5. Organic-rich shale of the Tuban Formation, Java, Indonesia.

fluids from a well, this gas can form bubbles. The bubbles start out small but grow in size as they travel upward through the well bore. This sets up an explosive chain reaction: As the bubbles grow they propel more oil out of the hole, thus releasing additional pressure and making the bubbles even bigger. The same basic principle can be demonstrated by shaking up a bottle of soda and is responsible for violent volcanic eruptions such as at Pompeii or Mt. St. Helens.

Oil well drillers have known about overpressure for a century or more and have developed a time-tested and generally reliable strategy for controlling it. As they are drilling they pump a slurry of water, clay, and other additives down through the inside of the drill pipe. After it exits the pipe this drilling mud returns to the surface through the space between the pipe and the hole (called an annulus). The additives allow the density of the drilling mud to be increased above that of water; that in turn increases the pressure the mud exerts within the hole. By adjusting the additives drillers can carefully balance the natural pressures encountered within reservoir rocks, preventing blowouts. Drilling mud also serves to cool the drill bit and to carry rock cuttings out of the hole.

Blowouts are now extremely rare events, but they can still occur. Most of these occurrences may be traced to simple human error or mechanical breakdowns. The infamous BP/Deepwater Horizon oil spill in the Gulf of Mexico in 2010 has been attributed to an unhappy coincidence of several such lapses, including the failure of cement that had been injected into the bottom of the well, the failure of emergency blowout preventers to seal off the wellhead at the seafloor, and the failure of anyone to notice early danger signs immediately prior to the blowout.

Timing Is Everything

The creation of a commercial oil or gas field is a geologically improbable event. A number of different geologic processes work together to form an active petroleum system (Figure 9.6), and commercial accumulations can only occur when they all fortuitously coincide. Exploration geologists have developed a habit of listing these processes individually in a sort of checklist, which must be satisfied before a successful well can be drilled. Each of the items on the list is potentially subject to uncertainty; if it does not perform as expected then failure is inevitable. The checklist looks something like this:

√ Source rock deposition
√ Reservoir (permeable) rock deposition
√ Seal (impermeable) rock deposition
√ Trap formation
√ Source rock burial
√ Migration of oil and gas

This is a rather long list, but it is still missing one very important item: timing. All of the events must not only have occurred, but have occurred in an appropriate sequence. It does no good to close the barn door after the horse has bolted. The petroleum geology equivalent to this expression would be to form a trap *after* oil migration has occurred. A number of other possible failure scenarios may be imagined, for example, depositing the seal before the reservoir (which reverses their required spatial relationship) or depositing the source rock after most of the subsidence of a sedimentary basin has already occurred (which prevents the source rock from being adequately heated). Another particularly important timing-related item is leakage from the trap. All oil and gas fields leak to some extent, because the seal rocks are by nature permeable, have been breached by fractures, or both. In the absence of ongoing oil generation and migration, the rate of leakage therefore limits a field's effective geologic lifetime.

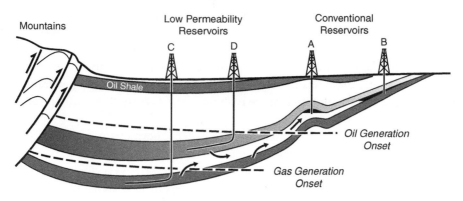

Figure 9.6. Schematic illustration of an active petroleum system.
Gray shading indicates organic-rich petroleum source rocks (mudstone). The
uppermost gray interval corresponds to oil shale, which requires artificial
heating to break down its complex organic matter into oil. Natural oil
generation begins approximately at the upper dashed line, and gas generation
becomes dominant below the lower dashed line. Conventional reservoirs
contain oil and gas that migrated out of source rocks and upward into a trap,
as illustrated by arrows leading to wells A and B. In contrast, wells C and D tap
oil and gas that are retained either within the source rocks or in other nearby
rocks with low permeability (see Chapter 10).

The actual geologic age of commercial oil and gas fields varies
widely. Some are relative newborns; for example, the Los Angeles Basin
in California only began to generate and trap significant quantities of
oil during the last 2–3 million years. At the other extreme are accumulations that persist for hundreds of millions of years after they first
formed. Fields in the Appalachian Basin, the birthplace of the U.S. petroleum industry, fall into this category. However, some of the most prolific provinces in the world correspond to sedimentary basins that span
long stretches of geologic time, with oil generation and entrapment
beginning tens of millions of years ago and continuing today.

The Gulf of Mexico is one such province, which traces its roots
back about 180 million years to the Middle Jurassic period, when
the supercontinent Pangea was in the process of dispersing into the
smaller continents we know today. Africa and South America started
drifting away from North America at this time, leaving behind a deep-
ening marine basin that was connected to the larger ocean by a relatively narrow strait. A warm climate caused intense evaporation of
the Gulf of Mexico, with this water continually being replenished

by inflow through the strait. The net result was buildup of several kilometers of bedded salt and limestone. Rivers draining the western mountains of Mexico and the United States soon began to dump mud and sand on top of the earlier deposits, eventually building the total sediment pile to thicknesses locally exceeding 16 km. Oil generation began from multiple source rock intervals as early as 40 million years ago and continues today.

This giant sedimentary pile exerts enormous pressure on the underlying Jurassic salt, forcing it to flow like toothpaste into massive underground pillars that sometimes pierce the land surface. Offshore, the seafloor is pockmarked by depressions formed above areas where salt has escaped. The entire deposit is continually collapsing under its own weight toward deeper areas of the Gulf of Mexico, folding and faulting its sedimentary layers in the process. These various disturbances create countless traps for oil and gas but also contribute to the leakiness of the system. For example, Prof. Lawrence Cathles of Cornell University estimated that in one area off the shore of Louisiana, about 70 percent of the oil and gas that have been generated has since leaked to the seafloor. Most of the remainder remains trapped in the source rocks or is actively migrating toward the seafloor, with only a small fraction being caught in traps.

Oil and Gas in the Persian Gulf Region

Nowhere on Earth have the various geologic elements required to make an oil or gas field combined in such spectacular fashion as in the Persian Gulf region (Figures 9.7 and 9.8). The lucky countries that directly abut this warm, shallow embayment together possess roughly half of the world's known reserves of conventional oil and 40 percent of its natural gas. No other region has anything close to this concentration of hydrocarbon riches. Why did geologic fortune smile so brightly on this otherwise desolate corner of the world? It is surprisingly difficult to point to any single geologic factor as overwhelmingly important or unusual. Rather, what sets the Persian Gulf and surrounding areas apart is the coincidence of a uniquely large *number* of favorable features, each of which alone might be considered unremarkable.

The Persian Gulf region does not possess the world's richest or thickest petroleum source rocks, but it does contain at least three to four major source intervals and many smaller ones that are each capable of generating impressive amounts of oil and gas. It does not possess uniquely porous reservoirs, but it does have at least a dozen major

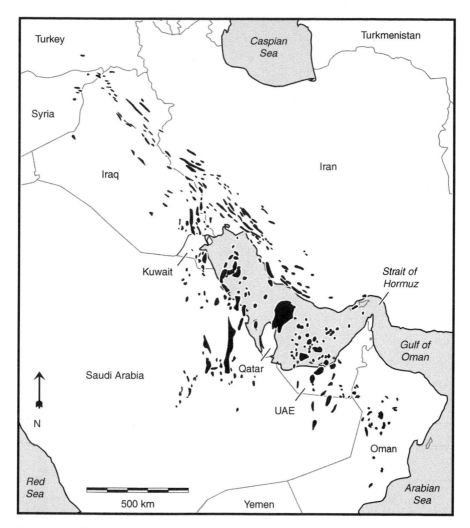

Figure 9.7. Oil and gas fields of the Persian Gulf province, shown in black (based on Middle East Oil and Gas Map, U.S. Central Intelligence Agency, 2007).

reservoir horizons and many more minor ones. It has no single giant salt horizon equivalent to the Jurassic salt in the Gulf of Mexico, but it has many smaller salt and mudstone beds that are effective in preventing the escape of hydrocarbons. Traps are numerous and large and have formed repeatedly through time. Oil generation began tens of millions of years ago and has continued through the present. In short, relentless geologic repetition, combined with prodigious amounts of

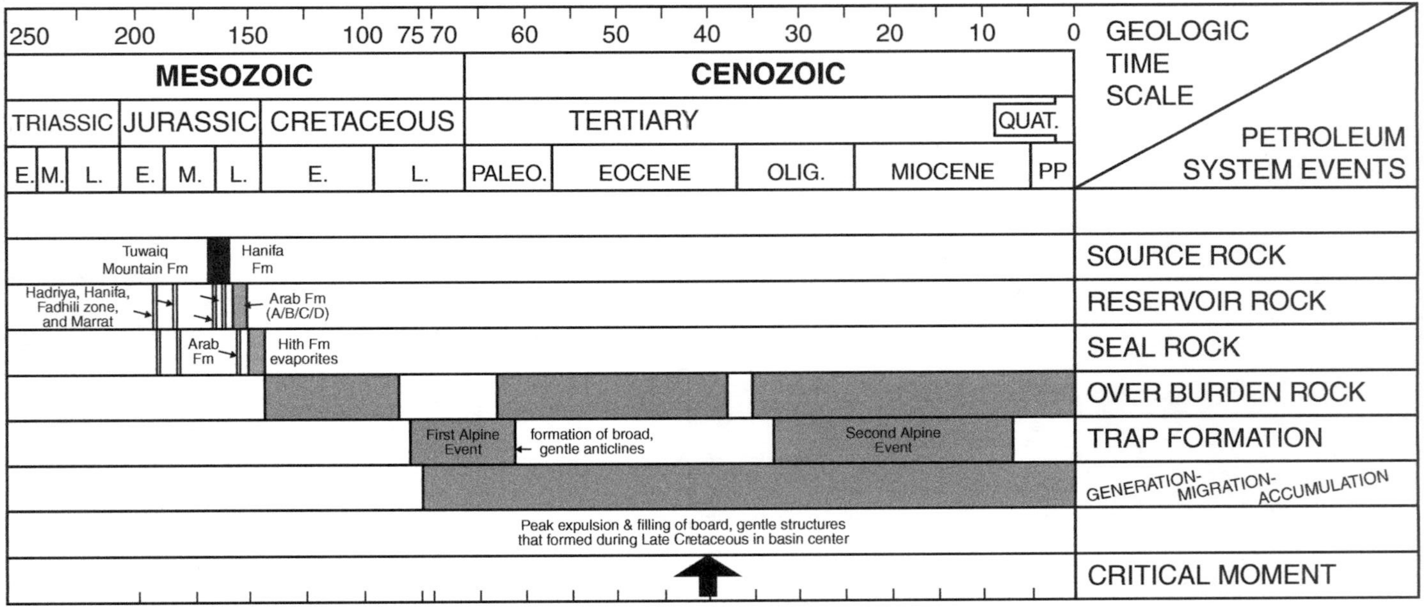

Figure 9.8. Timing of events related to the entrapment of oil from Jurassic source rocks in the Persian Gulf region (Pollastro, 2003).

luck, has helped to beat the odds that normally disfavor the formation of rich oil and gas fields.

For much of its geologic history the Persian Gulf region enjoyed a remarkably quiet existence, far removed from the major upheavals caused by the tectonic assembly and later dispersal of Pangea. An enormous shallow sea 2,000 km wide and 3,000 km long covered much of the region for much of the past 500 million years, gently laying down layer upon layer of sediment as sea level fluctuated up and down. No modern sea can really compare, but the Great Barrier Reef and associated northeast shelf of Australia probably are closest. During the Jurassic and Cretaceous periods the Persian Gulf basked in equatorial sunlight, which drove phytoplankton blooms, limestone precipitation, and evaporation, helping to create multiple source, reservoir, and seal horizons, respectively. Well-developed reefs like those of Australia were comparatively rare in this monotonous seascape, however. In general it was a spacious and quiet neighborhood ruled by long periods of stability, a perfect home for mud, sand, and salt.

Relief from this prolonged boredom occurred at last when northeast Africa (including what is now the Saudi Peninsula) crashed into Eurasia around 40 million years ago, heaving up the Zagros mountains (in what is now Iran and Iraq). Then about 25 million years ago the Red Sea opened, creating a new Arabian tectonic plate. Both of these cataclysms continue today, with the result that the Arabian Plate is being actively shoved into and beneath Eurasia. The weight of the rising Zagros has flexed the northeastern edge of the Arabian Plate downward, much as a diver would flex a diving board. This flexure made room for thick accumulations of sediment to be deposited; they in turn buried underlying source rock beds to depths where they began to generate oil and gas. Upward-migrating oil and gas permeated the Zagros mountains, filling numerous long, narrow anticlines that parallel the collision zone. Oil and gas also flowed long distances southwestward through rocks of the Arabian Plate, filling the giant anticlines of Saudi Arabia and surrounding countries.

The Persian Gulf petroleum province is clearly in its geologic prime, but it will not stay that way forever. Eventually the ongoing collision will run its course, existing oil fields will rupture and leak, and source rocks will reach exhaustion. None of this will happen soon, but 100 million years from now the picture will be very different. The largest conventional oil field in North America may provide an instructive glimpse of what is to come. Prudhoe Bay field in northern Alaska was once a proud member of a much larger family of oil and gas fields scattered across the arctic coastal plain. Widespread remnant traces of oil suggest that this area at one time contained reserves similar to the Persian Gulf today. Unfortunately for the United States, Alaska's golden moment arrived around 100 million years ago during the Cretaceous period, however. Its riches have been gradually slipping away ever since. Prudhoe Bay and associated fields on the coast of the Arctic Ocean are the last known remnants of this once gigantic petroleum province, which now holds only about one-fiftieth of the reserves of the Middle East.

A Needle in a Haystack: Finding Petroleum

A popular U.S. television comedy from the 1960s opened with a scene of a "poor mountaineer" discovering oil in the woods while he was out trying to shoot game to feed his family. The resultant financial windfall allowed him and his family to move away from the backwoods and into a mansion in Beverly Hills, California, with hilarious results. Although played for laughs on *The Beverly Hillbillies*, prior to the 20th century most oil was indeed discovered accidentally, with little or no understanding of its origins.

Geology entered the picture in a serious way only during the early 20th century, in response the need to locate ever-larger quantities of oil. The principal method for doing so was geologic field mapping, the primary goal of which was to locate anticlines that might potentially hold oil. If an anticline also featured some natural seepage, so much the better. Surface mapping led to the first major oil discovery in the Zagros mountains, in 1908 in what was then Persia (now Iran). A number of major surface anticlines were also discovered in other regions during the late 19th to early 20h centuries, including most of the larger discoveries in places such as California and Wyoming. Many of these anticlines are geologically obvious, impossible to miss if you know what you are looking for. Most can now be detected by the most elementary geology student with access to satellite imagery.

The supply of easily identifiable, undrilled anticlines had dwindled by the middle of the 20th century. The huge anticlines located on the desert plains adjacent to the Persian Gulf are poorly expressed at the surface and were initially overlooked. The major fields there thus were not discovered until the 1930s through 1950s. In addition to traditional field methods involving boots and rock hammers, geologists had to adopt new technologies, including measurement of subtle anomalies in natural gravity and magnetic fields that occur above anticlines. Eventually they began to use dynamite to generate sound waves, which reflect off subsurface structures and help to pinpoint them. The basic principle is the same as SONAR (SOund Navigation And Ranging), which is commonly used to detect concealed submarines or fish. In petroleum exploration it is known as reflection seismology, which is colloquially shortened to "seismic."

Geologists' understanding of how oil and gas fields form has grown more and more sophisticated, in parallel with improvements in technology, allowing clever detective work to gradually replace sheer luck in petroleum exploration. Intricately detailed three-dimensional seismic images of subsurface traps are now routinely obtained using computerized tomography techniques familiar to the medical imaging industry. In some cases it is possible to directly detect not only the structural outlines of subsurface traps, but even the fluid hydrocarbons that they contain.

A well drilled in an area with no previous oil or gas production has traditionally been referred in the industry as a "wildcat," implying a certain degree of reckless abandon and perhaps even fiscal irresponsibility on the part of the driller. Wildcat wells could be expected to succeed perhaps 10 percent of the time; they made better tax shelters than investment opportunities. True wildcats have become something of a rarity today, in part because a massive amount of drilling has already occurred in many prospective areas, and in part because subsurface imaging technology has removed much of the surprise factor. Exploration success rates are now higher than in the early days, despite the fact that the remaining traps are smaller and better hidden. This improved success rate is important because new oil and gas fields must bear not only the cost of their own discovery, but also the costs of any unsuccessful wells drilled along the way. Higher success rates therefore allow more expensive wells to be drilled, in turn allowing drilling in more remote and challenging environments that were previously inaccessible.

The net result of massive drilling and sophisticated technology is that most of the larger conventional oil and gas fields that will ever be found have already been found. There is little room for debate on this point; the size of new discoveries and amounts of oil and gas discovered have both been continuously declining globally since the 1980s. The hunt has now turned to deeper-water areas of the offshore continental margins and to onshore areas that might hold subtle traps that have somehow managed to escape earlier detection. The prey have generally become smaller and more elusive.

Squeezing the Sponge: Reserves Growth

In 1897 the author Mark Twain wrote to a friend that "the report of my death was an exaggeration," in response to a newspaper account that had questioned his health. The same might be said of many major oil fields, which have continued to produce for decades longer than expected. Such longevity requires that the ultimate production of oil and gas exceed originally estimated reserves, in apparent defiance of the laws of nature and economics. One especially dramatic example of this is the Midway-Sunset field in California, which was discovered in phases between 1890 and 1900. Its published reserves in the 1930s were a little less than 1 billion barrels. By 2000 it had already produced nearly 2.5 billion barrels, however, and its estimated ultimate recovery was nearing 3.5 billion barrels (Figure 9.9).

How can this be? One might suppose that once an oil field has been discovered it's a simple matter to measure its reserves, much as a gasoline station manager might check the contents of an underground supply tank with a dipstick. The level in the tank responds in predictable fashion to the amount of gasoline that has been pumped out of it, and when it gets low another delivery must be made or else the tank will run dry. Unfortunately this simple and intuitive analogy does not readily transfer to underground oil fields, because they do not behave the way simple storage tanks do. The stated reserves in many oil fields can and do increase without any new "deliveries" being made. However, these increases do not actually violate any laws of nature or (necessarily) involve shady accounting practices.

To understand reserves growth better it is helpful to consider a simplified version of the equation used universally for calculation of oil field reserves:

$$\text{Reserves} = \text{area} \times \text{thickness} \times \text{porosity} \times (1 - \text{water saturation}) \times \text{recovery factor}$$

Area in the equation refers to the lateral extent of the subsurface reservoir (as projected to the surface), and thickness its vertical extent that is saturated with hydrocarbons. Multiplied together these quantities give total reservoir volume. Multiplying by porosity gives the volume of open spaces where fluids can reside. The water saturation term deducts the fraction of pore space that is filled by water rather than hydrocarbons. Recovery factor is the fraction of oil and gas contained in the reservoir that can actually be brought to the surface.

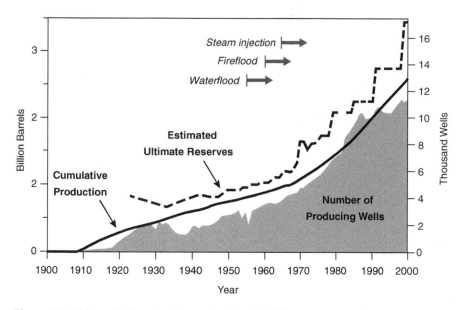

Figure 9.9. History of oil production and estimated ultimate reserves of Midway-Sunset field in California.

Estimated ultimate reserves includes past production plus remaining reserves. Estimated ultimate reserves more than tripled during the 20th century, as a result of increased drilling and the application of waterflood, fire flood, and steam injection techniques that boosted oil recovery factors. (modified from Tennyson, 2005 [USGS]).

Provided the input data are correct, this equation yields reserves numbers you can literally take to the bank, but the inputs seldom remain static. Instead they evolve continuously through time, as more wells are drilled and more oil and gas produced. The starting parameters are generally conservative, because reservoir engineers would rather underestimate reserves than overestimate them (financial regulators also frown on overestimation). Changes in the reserves calculation can be broadly cast into two types: revisions to the volume of pore space in the reservoir (defined by the first three terms of the equation) and changes in the recovery factor.

Reservoir Revisions

To apply the reserves equation literally a petroleum engineer must treat the reservoir as a relatively simple three-dimensional box of uniform porosity, or as a limited array of such simple shapes. The real world is far more complex, however. Suppose, for example, that the reservoir constitutes sandstone originally deposited by ancient rivers. Close examination of any modern riverbed will immediately reveal that it is not simply a linear sand box bordered by neatly rectangular walls. Real rivers contain countless sand bars of varying size, which change their shape and position continuously with every flood. Coarse sand and pebbles may be carried downstream during floods, whereas mud is deposited as floodwaters subside. Even the position of the river channel itself is not fixed in time. Some rivers continually shimmy sideways across their floodplain, eroding one bank and simultaneously building up sediment on the other. Other rivers consist of multiple, intertwined channels that shift position with each new flood. Rivers rarely stay in the same place for long unless interfered with by humans, and even then they chafe at the restraint.

It is simply impossible to map all of this complexity in the subsurface fully without drilling thousands of expensive wells, and even then many details would escape notice. Simplifying assumptions must therefore be made. As time progresses the reservoir typically becomes better understood, through monitoring of reservoir pressure and production histories and the drilling of more wells. The increased confidence that results may eventually lead to bolder estimates of reserves. Occasionally the opposite is true and reservoirs perform more poorly than expected, because of either overly optimistic projections or poor reservoir management. Either way, someone is not going to be happy.

An old adage among petroleum geologists and engineers holds that "the best place to drill for oil is in an oil field." In the early days of unregulated petroleum exploration this strategy often amounted to simple larceny, committed by bold opportunists who would attempt to drain a reservoir that someone else had already discovered. However, it is also common for new reservoirs to be legitimately discovered through deeper drilling below an existing field. Often the deeper reservoirs prove to be richer than the original shallow discoveries, because shallow oil accumulations may be degraded in quality by microorganisms (biodegradation) and other alteration processes. Such reservoir additions can substantially boost early estimates of oil field size.

Recovery Factor

Somewhat surprisingly, most of the oil ever discovered is still in the ground, even in the case of fields that have been largely depleted. Initial recovery factors may be as low as 5–10 percent, and rarely can

50 percent of the original oil be extracted. Depending on your point of view this represents either an inherent limitation or an opportunity. For example, boosting recovery from 10 percent to 20 percent in a particular field effectively doubles its reserves. The situation is a bit different for natural gas, which flows much more easily through small pores. The recovery factor for natural gas produced from conventional reservoirs may reach as high as 90 percent.

Oil recovery factors depend both on the mechanism by which oil is extracted and on the physical characteristics of the oil itself. In the early days recovery depended mostly on natural mechanisms that push oil from the reservoir and into a well bore. As noted in the earlier discussion of gushers, natural gas dissolved in oil provides one such mechanism. When reservoir pressure is released by the drilling of an oil well, gas can start to leave solution and form bubbles, which then expand and help push the oil toward the well bore. In other cases the gas segregates itself naturally before drilling ever occurs, forming a buoyant gas cap at the apex of a trap. This gas cap expands during oil production, helping to maintain reservoir pressure by pushing down on the underlying oil. In many oil fields, water provides an additional push from below, powered by regional pressure contrasts within an aquifer. Finally, oil can flow toward a well bore simply as a result of gravitational force if the well is pumped down to a level that lies below the original top of the oil in the reservoir.

These primary drive mechanisms may recover as much as 10–25 percent of the original oil in place. Obtaining greater recoveries generally requires secondary recovery efforts that involve injection of water, natural gas, or other fluids to maintain or restore reservoir pressures. Injecting water into the reservoir to maintain pressure during production is the most common and generally least expensive approach. Water injection, sometimes called waterflood, is the principal technique that has been used to extend production from the Midway-Sunset field and under favorable conditions can yield recovery factors as high as 35–45 percent. A significant downside to water injection is that some amount of water inevitably is produced with the oil. The magnitude of this "water cut" tends to increase through time, until further attempts at recovery eventually become uneconomical.

Enhanced Oil Recovery

Larry Lake, a renowned petroleum engineering professor at the University of Texas at Austin, wrote in 1992 that "traditional primary and secondary production methods typically recover one third of the oil in place, leaving two thirds behind." Lake and his coauthors from Chevron and Mobil noted that the remainder was left behind for the simple reason that it was cheaper to explore for new oil fields than it was to push the recovery factor higher in existing fields. At the time of that writing oil sold for a little more than $20 per barrel, which when corrected for inflation was fairly close to its average market value during the first seven decades of the 20th century.

By 2008 crude oil prices had passed $100 per barrel, and they have hovered near that level since. Exploring for new fields has simultaneously become more expensive. Under these conditions it becomes considerably more attractive to invest in recovery of oil from existing fields, using a variety of tertiary techniques that are loosely lumped together under the phrase "enhanced oil recovery" (EOR). One of the most effective techniques involves heating the oil in the reservoir, by either using steam or combusting some of the oil in place (known as fire flood). Heating dramatically reduces the viscosity of naturally thick, sticky oils, helping them to flow toward production wells. However, it also requires that significant amounts of energy must be expended in the process; for example, Lake and coauthors estimated that the energy equivalent of one barrel of oil must be expended to produce three to four barrels of new oil.

Various other approaches involve modifying the chemistry of water injected into the reservoir. For example, solvents such as liquefied petroleum gas, CO_2, nitrogen, flue gas, and alcohol can help oil to flow out of a reservoir. The addition of polymers acts to increase water's viscosity, in turn allowing it to displace viscous oil more effectively. Surfactants (soap) may be added to the water to help counteract the interfacial tension between water and oil, allowing the two to mix and flow out of the reservoir together. The underlying principle is similar to using a detergent to wash greasy dishes. Surfactants may also be produced directly within the reservoir by injecting a caustic base that reacts with the oil.

On the basis of data reported by the *Oil and Gas Journal*, EOR accounted for about 12 percent of U.S. oil production in 2010, with this percentage remaining fairly stable over the first decade of the 21st century (Moritis, 2010). This production resulted mostly from injection of steam or CO_2, with a lesser amount from injection of natural gas. Chemical treatments and other more exotic techniques such as the use of microbes to manufacture surfactants within the reservoir have made surprisingly little headway over the past twenty years.

Some of the most significant technological advances have resulted not from the development of new recovery methods, but from the use of information technology to optimize the application of existing techniques. Successful EOR depends on understanding the detailed interaction of many different variables, including the intrinsic geologic complexities of the reservoir and the physical and chemical behavior of several coexisting fluid phases. Sophisticated computer simulations that model the interplay of these variables can substantially improve oil recovery, although considerable uncertainty still remains. As of now EOR remains a relatively small contributor to world oil production.

Conclusions

The 20th century witnessed astonishing, unprecedented transformations to human society, brought about largely by the ready availability of crude oil and later by natural gas. These resources are concentrated by natural underground gathering systems, which collect oil and gas from a relatively large area of source rocks and focus it into much smaller traps. Finding these conventional oil and gas fields is often challenging, because of their comparatively small size and concealment beneath the Earth's surface. Once found, however, their freely flowing hydrocarbons can be easily extracted.

The richness of this prize has stimulated the development of extremely sophisticated technologies for discovering and producing oil and gas, and the search has progressively expanded into increasingly inaccessible realms. The cutting edge today lies offshore in water depths in excess of 3,000 meters and subsea drill depths greater than 10,000 meters. This trend cannot continue indefinitely, however, since petroleum source rocks are generally restricted to areas near the continents (see Chapter 7). Both the number and the size of new conventional oil and gas field discoveries have been declining for several decades. Future conventional oil and gas fields will likely continue to become smaller and harder to find, and more expensive to exploit.

In addition to the discovery of new fields, however, conventional oil and gas reserves in known fields may grow through improved development of existing reservoirs, and through drilling of new reservoir intervals that had not previously been tapped. Reserves can also grow through improvements in the percentage of the oil in-place that may be successfully extracted, particularly in the case of naturally viscous oil accumulations. Tracy Cook of the U.S. Geological Survey noted that between 2000 and 2009, reserves growth in existing U.S. oil fields outpaced newly discovered reserves. Such increases cannot continue indefinitely, but the precise limits of reserves growth are difficult to quantify.

For More Information

Attanasi, E. D., and Root, D. H., 1994, The enigma of oil and gas field growth: *American Association of Petroleum Geologists Bulletin*, v. 78, p. 321–332.

Beydoun, Z. R., 1998, Arabian Plate oil and gas: Why so rich and so prolific?: *Episodes*, 1998, v. 21, p. 74–81.

Cook, T. A., 2013, Reserves Growth of Oil and Gas Fields – Investigation and Applications: U.S. Geological Survey Scientific Investigations Report 2013-5063, 29 p.

Cathles, L. M., 2004, Hydrocarbon generation, migration, and venting in a portion of the offshore Louisiana Gulf of Mexico Basin: *The Leading Edge*, v. 23, 760–765, 770.

Demaison, G., and Huizinga, B. J., 1991, Genetic classification of petroleum systems: *American Association of Petroleum Geologists Bulletin*, v. 75, p. 1626–1643.

Houseknecht, D. W., and Bird, K. J., 2005, Oil and Gas Resources of the Arctic Alaska Petroleum Province: U.S. Geological Survey Professional Paper 1732-A, 11 p.

Hyne, N. J., 2012, *Nontechnical Guide to Petroleum Geology, Exploration, Drilling, and Production*: Tulsa, PennWell, 698 p.

Kvenvolden, K. A., and Cooper, C. K., 2003, Natural seepage of crude oil into the marine environment: *Geo-Marine Letters*, v. 23, p. 140–146.

Kvenvolden, K. A., and Rogers, 2005, B. W., Gaia's breath – global methane exhalations: *Marine and Petroleum Geology*, v. 22., p. 579–590.

Lake, L. W., Schmidt, R. L., and Venuto, P. B., 1992, A niche for enhanced oil recovery in the 1990s: *Oilfield Review*, v. 4, p. 55–61.

MacDonald, I. R., 1998, Natural oil spills: *Scientific American*, v. 279, p. 56–61.

Magoon, L. B., and Dow, W. G., 1994, The petroleum system, *in* Magoon, L. B., and Dow, W. G., eds., *The Petroleum System – from Source to Trap*: American Association of Petroleum Geologists Memoir 60, p. 3–24.

Moritis, G., 2010, Oil and Gas Journal , 2010, CO_2 miscible, steam dominate enhanced oil recovery processes: *Oil and Gas Journal*, v. 108, no. 14, p. 36–41.

Osborne, M. J., and Swarbrick, R. E., 1997, Mechanisms for generating overpressure in sedimentary basins: A reevaluation: *American Association of Petroleum Geologists Bulletin*, v. 81, p. 1023–1041.

Pollastro, R. M., 2003, Total petroleum systems of the Paleozoic and Jurassic, greater Ghawar uplift and adjoining provinces of central Saudi Arabia and northern Arabian-Persian Gulf: *United States Geological Survey Bulletin* 2202-H, 107 p.

Tennyson, M. E., 2005, Growth history of oil reserves in major California oil fields during the twentieth century: *U.S. Geological Survey Bulletin* 2172-H, 15 p.

10

Stuck in the Mud: Fossil Fuels That Fail to Flow

The oil and gas deposits that powered the 20th century can be elegantly summarized in a single word: mobility. Fluid hydrocarbons move freely through rock layers beneath the Earth's surface and under the right conditions can travel long distances to natural traps. Once trapped they readily flow to a well bore, making oil and gas relatively easy to transport to the surface. They can then be pumped across continents through pipelines or carried across the globe on tanker ships. Refined products flow easily though internal combustion engines, which have in turn revolutionized *human* mobility across the land, sea, or air.

The mobility of fluid hydrocarbons might be compared to vehicular traffic on a system of interstate highways or motorways. Just as high-speed roads transport people quickly between cities, certain porous and permeable rock types allow oil and gas to flow quickly (in geologic terms) toward underground traps. But what if there were no such subterranean superhighways? Or alternatively, what if the oil were simply too thick and sluggish to accelerate to highway speeds? Fluid hydrocarbons might then be forced to travel instead on the geologic equivalent of country lanes, gravel tracks, and muddy back roads, perhaps never reaching their destination at all.

In fact, much of the oil and natural gas that has been generated from source rocks in the Earth's crust has never moved very far. Some of it is retained within the source rocks themselves, which constitute less permeable, organic-rich mudstone, and some has migrated tens to hundreds of meters into adjacent beds that are only slightly more permeable. The relative immobility of such hydrocarbons means they do not become highly concentrated in localized traps; instead they remain spread out across large areas of sedimentary basins, just as coal is spread out across the Powder River Basin (see Chapter 8).

The U.S. Geological Survey describes these immobile oil and gas accumulations as "continuous," because of the observation that they lack the well-defined spatial boundaries imposed by conventional petroleum traps (see Chapter 9). The petroleum industry sometimes calls them "resource plays," although this term has no really precise definition. Descriptors such as "basin center," "shale," and "tight" are also commonly applied, adding more potential for confusion. All of the preceding terms, along with "coal bed methane," "heavy oil," and "methane hydrate," can be lumped under the heading "unconventional."

For most of the 20th century unconventional also meant uneconomical, and such deposits were simply ignored in assessments of commercial oil and gas reserves. The situation began to change near the dawn of the 21st century, however, because of rising fossil fuel prices and the wider application of horizontal drilling and hydraulic fracturing. Starting around 2005 natural gas and then oil production in the United States reversed their decades-long declines, and production began to *increase* at average annual rates of 3–5 percent. Most of this increase is from resources scarcely recognized to exist at the close of the 20th century.

Melting Pot: The Makings of Mudstone

The late Rodney Dangerfield built a comic career on his signature catchphrase "I don't get no respect." The same might well be said of mudstone, which is just compressed and hardened mud. Nothing evokes less emotion than common mud, and what feelings it does inspire are generally not favorable. Its more glamorous cousin, sand, seems to get all the attention. After all, who doesn't enjoy a visit to a sandy beach? Sand is macho; it soaked up the blood shed during gladiatorial contests in ancient Rome and is used today for sandblasting and to make concrete. Mud, on the other hand, is the annoying goo that sticks to your shoes or has to be hosed off the car. Mud is what's left behind after floodwaters have receded. Mud is also the most common type of sediment on Earth, a fact that does nothing to improve its image. The term "shale" sounds a bit more up-scale and frequently appears in the press as a synonym for mudstone, for example, in reference to "shale gas." Technically the term "shale" applies only to clay-rich mudstone that splits easily into thin parallel layers.

To the average person mudstone (or shale) would appear flabby and dull; it is often grayish in color, soft and crumbly in texture, and marked by unflattering horizontal stripes (for example, see Figures 7.8 and 9.5). Its dull appearance is deceiving, however. When viewed with a microscope, or better yet an electron microscope, mudstone reveals an almost shocking inner complexity. It is a geologic melting pot that amalgamates a wide variety of materials derived from the destruction of older rocks and from the demise of living organisms. Mudstone is in effect a microscopic mineral collection, zoo, and botanical garden all rolled into one.

The mineral collection can include dozens of different specimens, starting with silt grains of common rock-forming minerals such as quartz and feldspar (silt is the microscopic equivalent of sand). Even more important is clay, which is mostly produced by alteration of feldspar and other minerals during exposure to air and water (see Figure 2.6). The resultant clay minerals have a distinctive crystalline structure that resembles pages in a book. Their constituent atoms are arranged in two-dimensional sheets that are strongly bound within each sheet, but the individual sheets are only weakly bound to each other. Clay particles therefore tend to be platelike in shape, with lots of flat surface area.

Other common exhibits in the mineral collection include shards of volcanic glass (ash), the carbonate minerals calcite ($CaCO_3$) and dolomite ($CaMg(CO_3)_2$), submicroscopic quartz crystals, and noncrystalline silica (SiO_2). These constituents can achieve majority status in some mudstone types, which are named accordingly. For example, micrite is a mudstone made entirely from microcrystalline calcite, and diatomite consists entirely of the siliceous shells of diatoms (a type of phytoplankton). Other minerals form by precipitation within mud as it slowly transforms into rock. Calcite, dolomite, and quartz lead this list, which also includes various clay minerals, pyrite, siderite ($FeCO_3$), phosphorus-bearing minerals, manganese-bearing minerals, and others.

The organic remains entombed in mudstone constitute a cadaverous compendium of all five biologic kingdoms, ranging from primitive archaebacteria to whole skeletons of fish. Phytoplankton and other marine microorganisms usually dominate, but bits of wood, leaves, and spores, and pollen are also common. The rain of death to the seafloor or lake floor sustains an equally diverse population of scavengers, who burrow through the putrescent mud and leave behind their excrement (known as "coprolites" or "fecal pellets" in polite scientific company). Given an unlimited oxygen supply the scavengers would completely churn the sediment and consume all of the dead biomass,

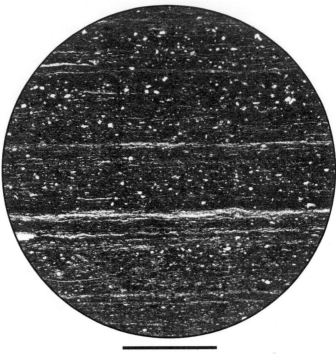

1 mm

Figure 10.1. Microscopic view of mudstone.

Note fine-scale horizontal lamination and the lack of visible pore spaces at this scale (compare to sandstone in Figure 9.2).

but bacterial decay often exhausts the oxygen supply before this can happen. Preserved organic matter may therefore exceed 20 percent of the weight of a mudstone.

Missing from the previous list is the single largest ingredient in freshly made mud, water. Like the human body, mud can contain as much as 75 percent water, and sometimes more. Water is gradually squeezed out as the mud becomes more deeply buried and compacted. Compaction contributes to the horizontal banding that is characteristic of mudstone and further obscures its cryptic texture. A comparison might be made to a birthday cake that has been flattened by a steamroller. You might perhaps be able to tell that it had been a cake, but the candles would be reduced to thin sheets of wax and any festive remarks or designs written in icing would become indecipherable. Lamination less than 1 mm thick commonly dominates the fabric of mudstone (Figure 10.1).

Failure to Launch: Oil and Gas Retained in Mudstone

Petroleum geologists have long known that oil and gas occur in diffuse form throughout sedimentary basins; gas in particular seems to be almost everywhere. In oil field language these small quantities are known as "shows" if encountered in a well bore and may provide a sign that something better (a major oil or gas field) is just around the corner.

Before oil and gas contribute to a conventional accumulation they must first leave their parent mudstone, a process fittingly termed expulsion. Once expelled, they become free agents that can migrate at will, either escaping to the surface or entering subsurface traps. This basic understanding was the cornerstone of petroleum geology for most of the 20the century. Every geologist knew that mudstone was not a reservoir, but rather a relatively impermeable seal that prevents oil and gas from escaping from a *real* reservoir. There was good evidence to support such a view: Wells drilled into reservoir rocks such as sandstone or limestone often produced strong and sustained flows of oil or gas, whereas wells that encountered only mudstone did not.

In hindsight this conventional wisdom was simply wrong. The clues were there all along; for example, drillers commonly encountered gas seeping from mudstone beds, even when a traditional reservoir rock had not been penetrated. Often the weight of the drilling mud had to be increased specifically to keep such seepage in check; otherwise it might contribute to collapse of the well bore or a blowout. Moreover, mudstone and other rock types normally considered to be impervious to hydrocarbon flow were also occasionally known to produce gas and oil, if they happened to be naturally fractured.

By the early to mid-2000s it was becoming apparent that impressive amounts of natural gas (and later oil) can often be coaxed out of mudstone and other low-permeability rocks. This new awareness quickly revolutionized the petroleum industry, but some important geological questions remain. For example, where exactly do the oil and gas reside in a mudstone? By what mechanism do they leave? How do we accurately predict and identify potentially productive horizons, assess their magnitude, and maximize extraction? Just as importantly, how will extraction of such resources impact groundwater, surface environments, and atmosphere?

Some of these questions are beginning to find answers. For example, scanning electron microscopy studies have shown that mudstone contains a surprising diversity of pores. Pores are found between

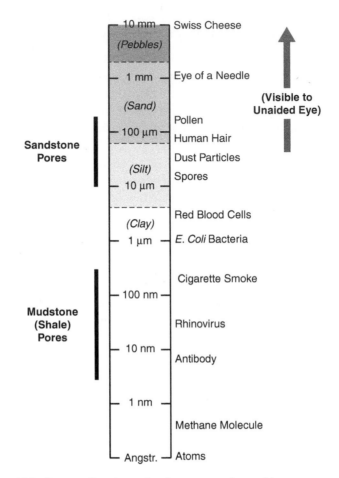

Figure 10.2. Comparative size scale of pore spaces in sandstone versus mudstone.

Note that each division on the scale represents a factor of 10 difference in size.

adjacent clay particles, within the microscopic shells left behind by phytoplankton, within microfractures, and even within fecal pellets! Some of the most common pores occur within bits of buried organic matter. In short, mudstone is full of holes, but the holes are very small, typically less than 100 nanometers (see Figures 10.2 and 10.3; a nanometer is *one-billionth* of a meter). At their smallest they approach the size of single methane molecules.

The magnitude of reserves held in a conventional oil or gas field can be readily calculated from first principles, using relatively

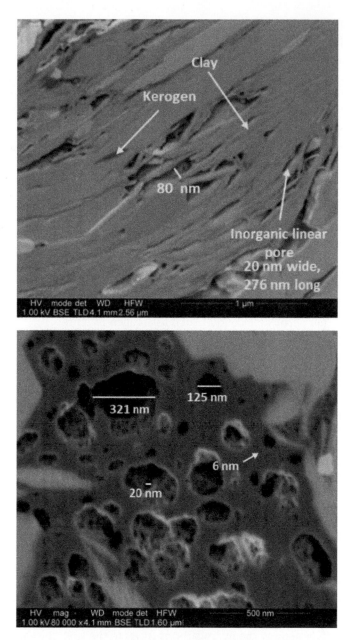

Figure 10.3. Scanning electron microscope images of nanoscale pores in mudstone.

Upper photo shows pore spaces preserved between adjacent inorganic mineral grains (clay). Lower photo shows pore developed in a particle of buried organic matter during transformation to fluid hydrocarbons (Curtis et al., 2010).

simple and long-accepted methods (see Chapter 9). The same cannot be said for mudstone reservoirs; no fundamental, unifying theory for predicting the productive potential of mudstone has yet emerged. Reserves estimates must instead rely on empirical data such as initial flow rates, and the rate at which production declines through time. At a larger scale, oil and gas productivity can be related to measurable rock properties such as mudstone thickness, mineralogy, organic matter content, and burial depth. These relationships can vary substantially from one area to another; local experience gained through drilling is therefore required to identify "sweet spots" or "fairways" of production.

Going to the Source: Horizontal Drilling

Oil and gas do not flow very quickly through mudstone; that much is clear. The key to extracting these substances is to reduce the distance through which they must flow before reaching either a well bore or some other permeable conduit such as an open fracture.

The worst-case scenario is a traditional vertical well that passes through a mudstone bed at or near a 90° angle. A vertical well might be compared to a single elevator that serves a very expansive office building, which measures thousands of meters on each side. Building occupants wanting to ride the elevator would first have to walk long distances to reach it, unless they happened to be stationed nearby.

Those farther away might choose instead to spend most of the day skulking in their offices, particularly if the building's hallways are narrow and serpentine rather than wide and straight. The natural pore networks through which oil and gas must flow might be compared to crooked hallways, which impede the ability of these fluids to reach a well bore.

Now imagine instead an elevator that descends to a particularly important floor, then turns to run horizontally down its longest dimension. Such a configuration would directly serve many more workers, encouraging them to leave their cubicles and jump on board. The oil and gas equivalent of this arrangement is a horizontal well. Rather than making the hydrocarbons travel to the well, the well travels to the hydrocarbons. The direction of a horizontal well can also be chosen to intersect naturally occurring vertical fractures, which act as feeder conduits to speed hydrocarbon flow.

Directional drilling has some other advantages as well; for example, wells can be drilled sideways underneath areas where surface access is impractical (such as towns or lakes). It also can greatly reduce the environmental impact of drilling by allowing multiple wells to be drilled from the same surface location. The sideways reach or "lateral" component of horizontal wells drilled in mudstone typically is in the range of 500–1,500 m. The record lateral distance achieved by directional drilling is about 10 km!

Directional drilling has a surprisingly long but not always proud history. The first recorded horizontal well was drilled in Texas in 1929, but the technology did not immediately result in commercial success, partly because of its increased costs compared to vertical drilling. At times directional drilling even carried an unsavory reputation due to its potential use for stealing oil from beneath another operator's lease. The most famous case of this may have occurred in the East Texas oil field in 1962–1963, when the Texas Rangers had to be called in to halt illegal slant drilling from old, unproductive well sites into competitors' adjoining leases that still held oil. Directional drilling achieved widespread acceptance in the 1980s, however, and has continuously grown in importance since, putting its checkered past behind it.

Cracking It Open: Hydraulic Fracturing

As useful as it is, horizontal drilling can still only access hydrocarbons that lie very close to the well bore, or that can flow there through natural fracture networks. Hydraulic fracturing (or hydrofracking) is designed to expedite matters, by augmenting natural fractures with an extensive network of new, artificial ones. This is done by pumping water into a well at pressures that exceed the failure strength of the reservoir rock. The basic principle is similar to blowing too much air into a balloon: Eventually the balloon will reach its elastic limit and rupture with a sudden bang.

In practice hydraulic fracturing is a bit more complicated. After a well has been drilled, a steel pipe called casing is installed and held in place by cement that is injected from the bottom of the well into the annular space between the casing and the hole. Then shaped explosive charges are used to perforate both the casing and the cement at multiple locations, creating entry points into the well bore. Next, the inside of the casing is sealed off around an interval that has been perforated, using donut-shaped rubber plugs called packers. Finally, water is pumped into the sealed section of casing until the pressure essentially shatters the rock, creating new fractures on the order of a millimeter in width. Usually this process is repeated at multiple points in the same well bore.

If only water were pumped into the well, the newly minted fractures would likely snap shut again as soon as the pressure was released. Sand (or another similar material) is therefore mixed with the injected water to prop the fractures open. The need for "proppant" has spawned a vigorous sand mining industry, and not just any sand will do. Ideal prop sand needs to have grains that are small enough to move into narrow fractures, are strong enough to stand up to high pressure without being crushed, and have nicely rounded edges that allow for easy fluid flow. Ironically some of the best frac sand in the United States is exported from Wisconsin, a state with absolutely no fossil fuel resources of its own. The same quartz-rich sandstone also serves as good raw material for making glass and photovoltaic cells.

Like horizontal drilling, hydraulic fracturing has a long history that predates the recognition that mudstone could be a reservoir. Hydraulic fracturing was first done in the late 1940s and its usage grew steadily in the decades thereafter; it has long been a common method for stimulating oil and gas production from marginal wells. It was used in an experimental mode to stimulate production of natural gas from mudstone as early as the 1970s, but found commercial success in the Barnett Formation of the Fort Worth Basin in Texas in the late 1990s to 2000s.

The Evolution of a Revolution

The Fort Worth Basin lies within an ecological region known as the Cross Timbers, a transitional mosaic of grassland, scrub, and diminutive oaks that originally separated the eastern Texas tall-grass prairies, savanna, and forests from the drier, mostly treeless plains to the west. Culturally it marks the edge of true cowboy country and was in fact crossed by cattle being driven north along the famous Chisholm Trail in the late 19th century. During the first half of the 20th century the Fort Worth sedimentary basin became a moderately prolific oil and gas producer, but by the 1960s there were no sizable fields left to find. Drillers were therefore forced to pursue smaller and less productive targets. Similar histories characterize many other onshore oil and gas producing areas of the United States, which were densely explored by the mid-20th century but gradually declining in production by its closing decades.

The decline of the Fort Worth Basin was eventually reversed by gas production from the Barnett Formation (commonly referred as the Barnett shale). Its organic-rich mudstone was deposited about 330 million years ago, in a narrow inland seaway. This seaway was originally bounded on the south by the equatorial mountain range that spanned the supercontinent Pangea (see Figure 8.1). The water was deepest near the mountains. Layers of organic-rich marine mudstone reached up to 250 m in thickness and were later buried by several kilometers of younger rocks. The entire area eventually rotated counterclockwise, so that the remnants of the ancient mountain chain now run northeast-southwest.

Extraction of natural gas from the Barnett Formation was hardly an overnight success. The discovery well was drilled in 1981, but its flow rates were only modest at best. A long period of experimentation ensued, during which many different combinations of hydraulic fracturing strategies and frac fluid recipes were tried and modified. Of particular importance was recognition that the density and distribution of natural fractures exerted a controlling influence on successful gas production. The first horizontal well in the Barnett was drilled in 1997, and within a few years horizontal drilling proved key to obtaining high production rates. Dramatic growth in Barnett Formation production began around 2000 and has continued since. Annual production increased fivefold from 2000 to 2005, and then tripled between 2005 and 2010. These increases were achieved through very dense development; by 2012 the Fort Worth Basin had been pincushioned by more than fifteen thousand wells, with the well sites spaced as close as 400 m (roughly a well every quarter-mile, or about two city blocks).

Success breeds imitation, and drillers quickly sought to reproduce elsewhere the same factors that worked so well in the Barnett. They were particularly successful in the Marcellus Formation of western Pennsylvania and adjoining areas. The Marcellus was deposited in the same inland sea as the Barnett, but about 60 million years earlier (see Figures 10.4 and 10.5). Both formations contain similar organic-rich mudstone deposits, although they differ somewhat in geologic detail. The Fort Worth Basin can boast of a somewhat higher gas concentration per acre, but the gas-bearing area of the Marcellus spreads out over a larger geographic area.

Mudstone formations have proven productive in a number of other similar sedimentary basins in the central and eastern United States and in geologically younger formations in the western United

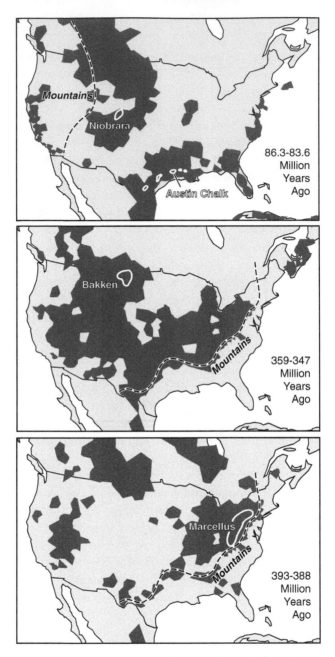

Figure 10.4. Flooding of the North American continent during selected time
intervals.

Dark gray indicates approximate areas where marine sedimentary rocks are
preserved. White outlines indicate areas of major oil or gas production from
mudstone or other low-permeability reservoirs, which were deposited during the
corresponding time interval (from the Macrostratigraphy Database, University of
Wisconsin-Madison, http://macrostrat.org; courtesy of Shanan Peters, Noel Heim,
and John Czaplewksi).

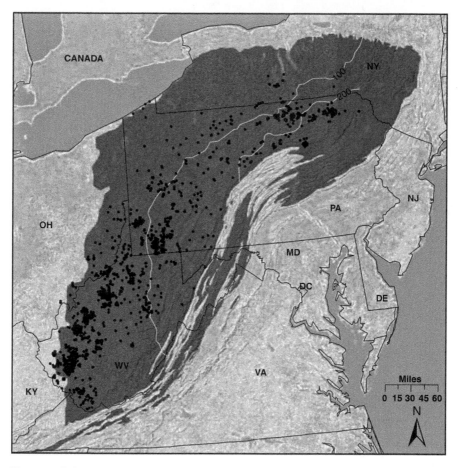

Figure 10.5. Extent of the Marcellus Formation in the eastern United States, shown in dark gray.

White lines indicate formation thickness in feet. Black circles indicate wells producing entirely or in part from the Marcellus Formation. (modified from U.S. Energy Information Administration).

States. According to the U.S. Energy Information Administration, total gas production from mudstone grew 241 percent between 2007 and 2009, and 257 percent between 2009 and 2011, easily outpacing every other domestic energy system. The size of the ultimately recoverable gas resources remains speculative, however. Wells drilled in mudstone flow at their highest rates immediately after completion. Production drops off dramatically in the first year or two, before eventually stabilizing in a longer, more gradual decline. Because of the relative youth of most existing wells, it is difficult to say how long they will remain commercially viable. Further complicating matters, it may be possible to rejuvenate productive areas by repeated hydraulic fracturing.

Tight as a Tombstone

In normal English usage the word "tight" can assume a remarkable variety of meanings, including close-fitting, stretched out, miserly, and even intoxicated. To a petroleum geologist it refers to a limestone or sandstone that proves to have low porosity or permeability. A tight reservoir was traditionally feared, because it meant a failed well. In really bad cases a geologist might colorfully lament that the reservoir had turned out to be as "as tight as a gnat's hatband" or "as tight as a tombstone." However, the same techniques that allow the production of oil and gas from mudstone have also been very successful in stimulating production from tight sandstone and limestone reservoirs.

One of the earlier tight reservoirs to be exploited was the Austin Chalk, a fine-grained limestone found to the south and east of Fort Worth. Outcrops of the Austin Chalk roughly parallel the Texas coastline (Figure 10.4) and mark the edge of its flat southeastern plains. It was deposited during the Cretaceous period, more than 200 million years after the Barnett Formation. The face of North America had changed considerably by this time; Africa had pulled away to open the north Atlantic Ocean, and the eastern mountains had worn down. A new series of ranges had risen in the west, within what are now the Rocky Mountains. An inland sea once again flooded the heart of the continent, but this time its deepest waters lay closer to the western mountains.

The Austin Chalk was deposited within the southern limits of this inland sea, in quiet waters relatively far removed from the mountains. It is composed largely of the microscopic, calcareous remains of one-celled phytoplankton called coccoliths, which also form the famous white cliffs of Dover. The Austin Chalk lacks the stark grandeur of its English cousin, however; it is much more modestly exposed in low escarpments, highway cuts, and river bottoms. These unassuming light-gray to yellowish outcrops typically possess only modest porosity and little intrinsic permeability and therefore might easily be written off as poor reservoir rocks. However, the chalk is by nature brittle and riddled with natural fractures that formed as it was gently flexed during burial. These fractures provide a natural plumbing system that helps oil and gas to escape and were the basis for a 1980s horizontal drilling boom in the Austin Chalk.

The Cretaceous seaway eventually began to fill in with river deposits, derived from the mountains to the west. The resultant land surface became home to countless generations of dinosaurs, who drank from the rivers and fed on the forested plains. The rivers deposited sand that eventually hardened into sandstone containing abundant natural gas. Extracting gas from these reservoirs is not easy, however. In some cases their pore spaces were partly filled with mud during original deposition. In other cases the natural plumbing between pores was clogged as warm waters circulated through the sandstone, precipitating minerals such as quartz, calcite, or clay. The same problem often plagues residential hot-water pipes in areas of hard water.

One of the more successful gas fields to produce from tight sandstone lies near the town of Pinedale, in western Wyoming. It is located within the Green River Basin, a rather desolate area of sagebrush-covered high prairie. Pinedale Field is defined by the Pinedale anticline, a cigar-shaped geologic structure that parallels the Wind River Mountains that tower to the east. Tourists hurrying north through this area on their way to the Teton and Yellowstone National Parks might notice the hundreds of oil wells to their left, but probably most prefer to gaze upon the beautiful snow-capped peaks to their right instead. Similar tight sandstone reservoirs occupy other sedimentary basins of the Rocky Mountain region, particularly in Wyoming, Colorado, and Utah. In 2011 tight gas accounted for approximately 25 percent of total U.S. production.

The Bakken Boom

Asked to produce a list of the top U.S. oil-producing regions, few people would think to include North Dakota, whose windswept high plains are home to more than twice as many cattle as people. In 2005 it ranked only tenth in oil production by U.S. states. By 2013 it had skyrocketed to #2 however, trailing only Texas. This newfound bounty was gained from horizontal drilling and hydraulic fracturing of the Bakken Formation (Figure 10.6). The development of this new resource has rapidly transformed the previously sleepy landscape of western North Dakota and caused the state's population to increase by 7.6 percent between 2010 and 2013.

The Bakken Formation is a close relative to both the Marcellus and Barnett Formations, deposited during marine flooding that penetrated into the heart of the continent (Figure 10.4). Unlike the

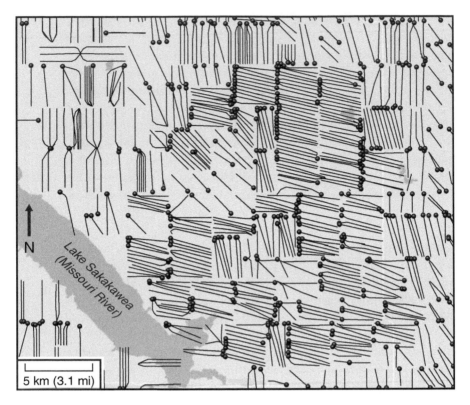

Figure 10.6. Map of wellhead locations (circles) and horizontal well bores (lines) drilled into part of the Bakken Formation, North Dakota (well bores projected to surface).

The roughly circular area of high-density drilling is the Sanish Field, one of the most prolific areas of Bakken Formation oil production. The target interval lies more than 3,000 m below the surface, a distance approximately equal to the maximum width of the segment of Lake Sakakawea shown in the lower left (data obtained from the North Dakota Industrial Commission and plotted through www.fractracker.org).

Barnett or Marcellus, the Bakken Formation was deposited in a relatively gentle, dish-shaped depression and was less deeply buried. Its organic matter has generated mostly oil instead of gas. Some of this oil migrated relatively long distances (kilometers to tens of kilometers) into conventional petroleum reservoirs of the Williston Basin, which covers parts of North Dakota, Montana, and the Canadian provinces of Manitoba and Saskatchewan. Relatively small oil fields have been commercially exploited there for many decades. The majority of the oil never traveled very far away from its source rock, however, migrating perhaps a few hundreds of meters at most.

The Bakken Formation reservoir is sometimes likened to an Oreo cookie, because two dark-colored organic-rich mudstone layers sandwich a sandstone layer in between them. Press accounts commonly label the contents of this reservoir as "shale oil," but this is misleading. It is true that some oil is extracted from fractures in the mudstone beds, but the majority actually is from the tight sandstone and adjacent tight limestone bed. The term *shale oil* also creates confusion because of its similarity to *oil shale*, which is an entirely different animal. Whereas the Bakken Formation contains oil that has already been generated, *oil shale* is a rock that must be artificially heated to break down its complex organic polymers (see Chapter 8). A better term in the case of the Bakken Formation would be "tight oil" (note that it is the reservoir that is considered tight, not the oil itself!).

In 1995 the U.S. Geological Survey estimated that the Bakken held only 151 million barrels of recoverable oil, but by 2008 they revised their assessment to 3.65 *billion* barrels, a 24-fold increase. In 2013 they increased their estimate again, concluding that 7.4 billion barrels of oil are recoverable from the Bakken and underlying Three Forks Formation combined. This total equals about half the originally recoverable reserves of the Prudhoe Bay field in Alaska, the largest oil field in the United States. Similar tight oil reservoirs are rapidly being developed in other areas of the United States and in total appear to be quite large in magnitude. Some of the more prominent examples include the Cretaceous Niobrara Formation in northeastern Colorado and the Eagleford Formation in south Texas, both of which are close geologic relatives of the Austin Chalk (Figure 10.4).

Frac Impacts

The recent boom in North Dakota is emblematic of numerous other areas containing newfound fossil fuel resources. Small towns became boomtowns, multiplying their populations many times. The arrival of thousands of new workers created serious traffic problems in places where the installation of a new stoplight might previously have made the front page of the weekly gazette. This was all good news for motels, discount retailers, and clown-themed fast food restaurants, but it must also have been jarring to the original residents. Ironically, many of the areas caught up in the recent growth of oil and gas production have seen similar boom times before, but the older oil and gas fields have long since declined. The new drilling boom has also been much more expansive than more traditional oil and gas developments, covering whole counties with densely spaced roads, drill pads, and pipelines.

Increased drilling has also spawned an increasingly virulent public backlash, focused on its potential environmental consequences. A wide variety of concerns have been voiced, ranging from obvious problems of land surface disruption and air pollution to the suspicion that hydraulic fracturing might actually cause damaging earthquakes. Perhaps most concerning is the possibility that oil and gas development could threaten freshwater supplies that are needed for agricultural and residential use. According to a U.S. Geological Survey study of drilling in the Marcellus Formation, hydraulic fracturing of a single well can easily consume 11–19 million liters of water, roughly the equal of five to eight Olympic swimming pools. Even more worrisome is the eventual fate of that water, which contains a toxic mix of chemicals intended to enhance the effectiveness of hydraulic fractures (viscosity enhancers, surfactants, emulsifiers, biocides, and others).

The exact recipes for various frac fluids are closely held trade secrets, developed through trial and error by individual companies seeking competitive advantage. It is clear that at least some of the ingredients are unhealthy for humans, however. A 2011 U.S. congressional report, for example, found that methanol and ethylene glycol are commonly added to frac fluids, along with various other toxins and carcinogens. While the percentage of these additives is commonly less than 1 percent of the total fluid volume, 1 percent of 19 million liters would still fill several backyard swimming pools. Moreover, should frac fluids somehow find their way into the local water supply it would hardly be comforting to learn that your drinking water only contains "a little" poison!

But can frac fluids really enter the freshwater supply? This question is hotly debated, and actually there are two separate questions to consider. The first part concerns the fate of water that flows through a well up to the surface, which includes both flow-back water left over from hydraulic fracturing and natural brines contained in the reservoir. These large volumes of toxic frac fluid and brine could potentially contaminate fresh surface waters if not carefully handled. They must be treated to remove their harmful constituents, recycled for new hydraulic fracturing, or injected through disposal wells drilled into deeper reservoirs. Done right, none of these practices necessarily poses any long-term health hazard. They do add cost and complexity, however, and the chance that something could go wrong.

The second part is more controversial: Can hydraulic fracturing itself directly contaminate freshwater aquifers in the subsurface? Early public perception of this question was shaped by viral videos showing domestic faucets that simultaneously flowed both water and methane, the latter in flames. Although methane and other light hydrocarbon gases are not toxic per se, asphyxiation, fire, and explosion are obvious concerns. There have also been anecdotal reports of well water contamination by frac fluids, although few systematic studies have yet appeared. The idea that hydraulic fracturing intrinsically causes groundwater contamination remains unproven.

Overinflated Fractures?

If hydraulically induced fractures accidentally extend beyond the targeted oil or gas reservoir, could they provide a pathway for contaminants to flow into shallower aquifers? The geologic odds appear to heavily disfavor such an occurrence. Most water wells reach depths of a few hundreds of meters at most. In contrast, hydraulically fractured gas wells are usually drilled to depths of 1,500 m or more (Figure 10.7). Drillers have strong financial incentives not to project hydraulic fractures above the target reservoir, because of the added cost in frac fluid, proppant, and energy. Taller-than-needed fractures could also penetrate into adjacent permeable beds, which might absorb fluid and reduce the effectiveness of the frac job. Such zones might also allow additional brine to flow into the well, which would then require expensive disposal.

Fortunately the basic mechanics of hydraulic fracturing tends to prevent it from extending farther than intended. A balloon analogy partly illustrates why. A typical party balloon can be easily inflated to its bursting point using only lung power. However, if you attach the same balloon to a length of hose and immerse it under 1 m of water it will become nearly impossible to inflate, much less burst, because of the water pressure exerted on it. Your diaphragm must effectively lift the entire column of water that lies above the balloon. If the balloon is, say, 20 cm in diameter, this would be the equivalent of trying to breathe while lying on your back with a 40 kg weight placed on your chest.

Hydraulic fracturing likewise requires large pumping pressures (in the range of hundreds of times atmospheric pressure) to overcome the pressure exerted by the weight of overlying rocks. Once fracturing begins, frac fluid flows from the well bore into the newly made

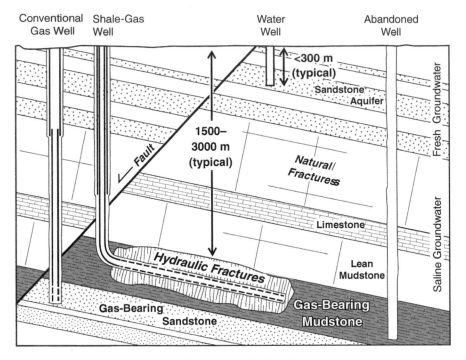

Figure 10.7. Schematic illustration of hydraulic fractures in relation to freshwater aquifers.

Note that hydraulic fractures themselves are unlikely to penetrate shallow aquifers. Contamination of aquifers by natural gas, oil, or saline waters could potentially occur through faults or other natural fractures intersected by hydraulic fractures, although such occurrences are not common. Leakage could also occur via abandoned wells that have not been properly sealed, or via wells in which steel casing has not been properly cemented in place.

spaces in the rocks. Increasing amounts of fluid must be pumped to counteract the weight of overlying rocks and expand the fractures; runaway fractures are therefore very unlikely. Fracture propagation largely ceases once pumping stops.

Runaway fractures are further inhibited by the complex interlayering of different sedimentary rock types, which have varying degrees of fracture resistance. At the shallower levels associated with freshwater aquifers, fractures also become more likely to propagate horizontally rather than vertically, as the weight of overlying rocks decreases.

The actual distribution of hydraulic fractures for a given well can be directly imaged in three dimensions by recording the tiny earthquakes they cause (called microseismicity). The size of these microquakes is tiny; Professor Mark Zoback of Stanford University and colleagues pointed out in a 2010 report that the seismic energy released by a single hydraulic fracture is about the equivalent of a gallon of milk dropped from chest height to the floor. They are not noticeable on the surface but can be detected by microphones placed in an adjacent well. Imaging of microearthquakes has revealed that induced fractures rarely travel more than a few hundred meters upward from their associated horizontal well bore.

Flaming Faucets: Gas and Groundwater

Minor quantities of methane have long been known to occur in groundwater; it is, after all, a very simple and ubiquitous substance. Methane found at depths of more than a kilometer or two mostly results from the thermal breakdown of sedimentary organic matter and is therefore termed thermogenic. Most of the gas contained in mudstone and other low-permeability reservoirs is of the thermogenic variety. Methane can also be produced by certain microorganisms called methanogens, which commonly live in swamps, landfills, shallow aquifers, and our own digestive systems. This biogenic methane is easily distinguishable from its thermogenic cousin on the basis of carbon isotopic ratios. Biogenic methane is normally unrelated to commercial natural gas production.

Thermogenic methane generated within deep sedimentary basins naturally migrates upward. In the absence of thick, relatively impermeable horizons some of it will likely find its way into shallower aquifers. Its mere presence in an area of intensive drilling and hydraulic fracturing therefore does not constitute proof of artificial contamination. However, recent studies led by Stephen Osborn and colleagues from Duke University reported that in areas of active gas extraction in Pennsylvania, methane concentration in water wells increases with proximity to the nearest gas well. They also showed that this additional gas was thermogenic and concluded that commercial gas extraction was indeed responsible for increasing the methane content of drinking water. They failed to find evidence of contamination of aquifers by frac fluids or brines, however, and the actual mechanism by which natural gas entered the water wells remains in doubt.

A subsequent 2013 study sponsored by Cabot Oil and Gas Corporation and published in the scientific journal *Groundwater* challenged the conclusion that methane in groundwater was related to nearby drilling. Its authors instead proposed that aquifer gas concentrations were governed by topography, with the greatest concentrations found in valley bottoms. The valleys in this region tend to form where natural fractures have promoted faster erosion; these same fractures could therefore enhance the flow of groundwater and possibly natural gas. The authors further argued that the gas did not even originate in the commercially productive Marcellus Formation, but in younger rock layers instead.

Similar controversies have erupted in other regions, with one of the more notable revolving around the Pavillion gas field in Wyoming. A draft report of the U.S. Environmental Protection Agency (EPA) released in 2011 suggested that a variety of inorganic and organic chemicals commonly used in frac fluids had contaminated deep, freshwater aquifers in this area. The EPA also noted that methane concentrations in shallower domestic water wells appeared to increase in proximity with gas production wells. In one case the concentrations were so high that they appear to have caused a shallow domestic water well to blow out during drilling, no doubt an unpleasant shock to the drillers!

Reading more deeply, however, it might actually seem surprising if the Pavillion gas field did *not* have problems with groundwater contamination. The gas wells are relatively shallow, with production occurring as close as 372 m to the surface, and the deepest water wells in this semiarid region penetrate as far as 244 m. The productive gas interval is therefore unusually close to freshwater aquifers. Worse still, the local geology tends to blur any true distinction between the two.

Pavillion field is located in the Wind River sedimentary basin, which 50 million years ago was a closed depression walled in by mountains on all sides. Rivers carried mud, sand, and gravel into the basin, depositing them in an ever-changing mosaic of channels, bars, and floodplains. This resulted in a complex array of stacked, discontinuous sandstone reservoirs, which are separated by multiple, discontinuous intervals of mudstone. There is, however, no single, impermeable sealing layer that might effectively inhibit the upward flow of gas. It is a fundamentally leaky sedimentary basin, with the same geological formation hosting both gas and freshwater.

The draft EPA report was quickly challenged by the State of Wyoming and Encana, the field operator. Both pointed to the complex and leaky nature of the basin to explain the occurrence of methane in the deep monitoring wells. They also claimed that the EPA report was inherently flawed and that the reported contaminant chemicals could have been introduced during the drilling of the EPA's own monitoring wells.

Perhaps the clearest conclusion to be drawn from the preceding controversies is that local geological conditions are paramount in determining the potential for groundwater contamination related to hydraulic fracturing. The nature of gas-bearing rocks, aquifers, and their relationship to each other vary greatly in different geographic areas. It seems highly unlikely therefore that any "one size fits all" answer to these questions can ever be achieved.

More mundane technical problems related to the drilling and completion of oil and gas wells can also result in groundwater contamination. Freshwater aquifers are normally protected from contamination by steel casing that is cemented in place before hydraulic fracturing commences. The cementing process is not foolproof, however, and it is possible for an imperfect cement job to allow leakage of contaminants through the space between the casing and the drill hole. Contamination can also arise from inactive wells that have been improperly plugged and therefore serve as open conduits between deep rock layers and shallow aquifers. Western Pennsylvania, for example, was the location of the first commercial oil well in the United States, drilled in 1859, and remained the center of the American oil industry for decades thereafter. Well abandonment procedures in those early days were irregular at best, and it would be virtually impossible to locate every old well bore.

Most of these problems are not unique to either horizontal wells or hydraulic fracturing. They can also occur with conventional wells. Routine mishaps can be expected to occur with drilling generally (usually at a low rate), just as they do with any large-scale industrial process. To some extent they are therefore a predictable consequence of our desire for cheap energy.

Getaway Gas: Accidental Methane Emission

Methane (CH_4) and carbon dioxide (CO_2) are both greenhouse gases that contain carbon, but CH_4 is much more potent in terms of global warming potential. Methane enters the atmosphere naturally from a variety of sources, ranging from natural leakage of underground gas accumulations to flatulent cows. It can also enter the atmosphere accidentally during well drilling, during gas transmission through pipelines, or even from leaky residential supply lines. Such accidental leakages are commonly labeled "fugitive emissions," since the gas has escaped its intended confinement.

Methane can also escape from the flow-back fluids released during hydraulic fracturing. If large enough, such losses could have a major impact on global warming potential. A group of Cornell University researchers claimed in a 2011 study that these emissions are indeed large. They estimated that between 3.6 percent and 7.9 percent of the methane produced using hydraulic fracturing escapes into the atmosphere, and that these escape rates are 30 percent to 100 percent greater than those associated with conventional gas production. They therefore concluded that the overall global warming footprint of burning natural gas extracted from mudstone is at least 20 percent worse than if the same energy were obtained from burning coal!

The Cornell study generated a swift and severe backlash from several other groups, including researchers at Argonne National Laboratories, Carnegie Mellon University, Massachusetts Institute of Technology, and even a different group at Cornell. One of their principal objections was that the release of methane from flow-back fluids was greatly overestimated. In practice the majority of this gas is either flared, producing much less harmful CO_2, or else captured for sale. The physical basis on which the global warming potentials of methane and CO_2 were compared was also questioned, and all of these groups agreed that the global warming footprint of gas extracted using hydraulic fracturing is substantially less than that of coal. As of this writing the controversy has not ended, however, and new studies are appearing regularly.

Conclusions

The past two decades have witnessed a sea change in our perception of fluid fossil fuel. For most of the past century oil and gas extraction relied on localized resources that were often challenging to discover, but relatively easy to extract. In contrast, 21st century oil and gas production has increasingly focused on accumulations that are easier to find, but more difficult to produce economically. This transition might be usefully illustrated through analogy to an apple tree. In the 20th century we picked the low-hanging fruit. Exploration efforts focused on the search for more trees, hidden away in a multitude of obscure locations, from which more low-hanging fruit could be picked. However, the new focus is on picking the apples that hang higher up in the tree. To do this requires using a longer ladder, or perhaps shaking the tree to dislodge its apples. Either approach adds difficulty, energy, and expense to the endeavor. The central question has therefore changed from "how many apples can we find?" to "How many apples can we afford to pick?"

Note that no distinct line can be drawn to indicate where the low-hanging fruit ends, and where the rest of the apples begin. Likewise, no clear natural line separates conventional from unconventional oil and gas reservoirs. Natural permeability in sediment and sedimentary rocks varies across an extremely wide range, encompassing twelve orders of magnitude. This suggests that extraction costs and technology, rather than the natural abundance of oil and gas, will ultimately determine what can be economically produced. This determination is by nature flexible, varying with changing technology and resource price.

Exploitation of low-permeability reservoirs is clearly more invasive than conventional oil and gas field development, however, because it requires closer contact with more rock. The traditional image of the lone wildcat well, drilled in hopes of discovering hidden treasure, is rapidly giving way to a more industrial scene, featuring row upon row of nearly identical wells designed to minimize cost. There is presently little scientific evidence that this new mode of extraction poses inherently new or unique environmental dangers. However, increasing the magnitude and density of oil and gas extraction will inevitably increase exposure to already familiar problems.

For More Information

Anonymous, 2013, Modern Shale Gas Development in the United States: An Update: National Energy Technology Laboratory, 79 p.

Bruner, K. R., and Smosna, R., 2011, A Comparative Study of the Mississippian Barnett Shale, Fort Worth Basin, and Devonian Marcellus Shale, Appalachian Basin: *U.S. Department of Energy/National Energy Technology Laboratory Report* 2011/1478, 106 pp.

Cathles, L. M., Brown, L., Taam, M., and Hunter, A., 2012, A commentary on "The greenhouse-gas footprint of natural gas in shale formations" by R. W. Howarth: *Climatic Change*, v. 113, p. 525–535.

Curtis, M. E., Ambrose, R. J., Sondergeld, C. H., and Rai, C. S., 2010, Structural Characterization of Gas Shales on the Micro- and Nano-Scales: Canadian Society for Unconventional Gas/Society of Petroleum Engineers Conference Paper 137693, 15 p.

DiGiulio, D. C., Wilkin, R. T., Miller, C., and Oberley, G., 2011, Investigation of Ground Water Contamination Near Pavillion, Wyoming: *EPA Draft Report* 600/R-00/000, 121 p.

Duggan-Haas, D., Ross, R. M., and Allmon, W. D., with Cronin, K. E., Smrecak, T. A., and Auer Perry, S., 2013, *The Science beneath the Surface: A Very Short Guide* to the Marcellus Shale: Paleontological Research Institution Special Publication 43, Ithaca, New York, 252 p.

Howarth, R. W., Santoro, R. W., and Ingraffea, A., 2011, Methane and the greenhouse-gas footprint of natural gas from shale formations: Climatic Change, DOI: 10.1007/s10584-011-0061-5.

Loucks, R. G., and Ruppel, S. C., 2007, Mississippian Barnett Shale: Lithofacies and depositional setting of a deep-water shale-gas succession in the Fort Worth Basin, Texas: American Association of Petroleum Geologists Bulletin, v. 91, p. 579–601.

Loucks, R. G., Reed, R. M., Ruppel, S. C., and Hammes, U., 2012, Spectrum of pore types and networks in mudrocks and a descriptive classification for matrix-related mudrock pores: American Association of Petroleum Geologists Bulletin, v. 96, p. 1071–1098.

Molofsky, L. J., Connor, J. A., Wylie, A. S., Wagner, T., and Farhat, S. K., 2013, Evaluation of methane sources in groundwater in northeastern Pennsylvania: Groundwater, v. 51, p. 333–349.

Murray, G. H. Jr., 1968, Quantitative fracture study – Sanish Pool, McKenzie County, North Dakota: American Association of Petroleum Geologists Bulletin, v. 52, p. 57–65.

Osborn, S. G., Vengosh, A., Warner, N. R., and Jackson, R. B., 2011, Methane contamination of drinking water accompanying gas-well drilling and hydraulic fracturing: Proceedings of the National Academy of Sciences, v. 108, p. 8172–8176.

O'Sullivan, F., and Paltsev, S., 2012, Shale gas production: potential vs. actual greenhouse gas emissions: Environmental Research Letters, v. 7, 6 p.

Pollastro, R. M., Cook, T. A., Roberts, L. N. R., Schenk, C. J., Lewan, M. D., Anna, L. O., Gaswirth, S. B., Lillis, P. G., Klett, T. R., and Charpentier, R. R., 2008, Assessment of undiscovered oil resources in the Devonian-Mississippian Bakken Formation, Williston Basin Province, Montana and North Dakota, 2008: U.S. Geological Survey Fact Sheet 2008–3021, 2 p.

Potter, P. E., Maynard, J. B., and Depetris, P. J., 2005, Mud and Mudstones – Introduction and Overview: Berlin, Springer-Verlag, 297 p.

Price, L. C., and LeFever, J. A., 1992, Does Bakken Horizontal Drilling Imply a Huge Oil-Resource Base in Fractured Shales?, in Schmoker, J. W., and Coalson, E. B., and Brown, C. A., eds, Geological Studies Relevant to Horizontal Drilling: Examples from Western North America: Denver, CO, Rocky Mountain Association of Geologists, p. 199–214.

Zoback, M., Kitasei, S., and Copithorne, B., 2010, Addressing the Environmental Risks from Shale Gas Development: Washington, D.C., Worldwatch Institute Briefing Paper, 18 p.

11

Petrified Petroleum: Oil Sand and Gas Hydrate

The preceding chapters described mobile, naturally concentrated hydrocarbon fuels (Chapter 9) and relatively immobile hydrocarbons that remain "stuck" near their source (Chapter 10). There is yet a third category that must be added to complete this list: oil and gas that successfully escape their source beds but then lose their mobility as they approach the Earth's surface. They now lie imprisoned, condemned to a monotonous and indefinite existence in one of two forms: thickened crude oil that has ceased to flow, or natural gas frozen into an exotic form of ice. Superficially these two could not appear more different; thickened crude is sticky and dark, whereas gas hydrate is crumbly and white. However, they share an acquired immobility. Rather than flowing through their host rock these deposits have become a part of it. They may reasonably be considered "petrified" oil and gas.

Thickened crude oil and gas hydrate both owe their existence to the intervention of microorganisms. Remarkable as it may seem, the petrification of crude oil results primarily from scavenging by specialized microbes that eat hydrocarbons. These microbes fall within two different groups of organisms that evolved billions of years ago: bacteria and archaea. The first group breathes oxygen (or sulfate, SO_4^{2-}) and is responsible for most of the decay that normally inhibits the geologic preservation of dead organic matter (see Chapter 7). The second group includes methanogens, microbes that live off the organic remains left behind after bacterial decay. They earn their name by emitting methane as a by-product. Unlike the aerobic bacteria that started the job, methanogens operate entirely without oxygen and in fact cannot tolerate its presence. They are among the earliest inhabitants of Earth, having evolved before the atmosphere contained free oxygen. By necessity they now hide from the light of day, buried in oxygen-depleted sediment beneath mires or the ocean, or within our own lower digestive tracts.

These microbial specialists can only digest (or "biodegrade") relatively small, light molecules. They preferentially consume the natural solvent that gives oil its mobility and leave behind an indigestible goop, that clings tenaciously to mineral surfaces and binds together sand grains or other loose sediment. Recent studies suggest that methanogens do most of the damage to oils contained within subsurface reservoirs, with bacteria playing a subordinate role. In addition to biodegradation, evaporation helps remove the more volatile compounds from crude oil, and contact with moving groundwater helps remove compounds that are water-soluble.

Conversely, microbial feasting is just the beginning of the journey for most of the gas that ends up in gas hydrates. Once generated, biogenic methane can migrate freely into more porous reservoir rocks, in a manner identical to gas generated by geological heating. Left to their own devices both types of gas would eventually continue to seep upward to the land surface or sea bottom and be lost to the atmosphere. Both types of gas can also be frozen in place before this happens, however, if they chance upon the right combination of low temperature and high pressure within water-saturated sediment. Such conditions are commonly found beneath arctic permafrost and below much of the seafloor adjacent to the continents. The result is the formation of tiny, three-dimensional cages of ice that imprison gas molecules.

In its usual sense the word "petrified" refers to dinosaur bones, tree trunks, or other once-living things that have been turned into stone. This process is clearly permanent, but the petrification of oil and gas can be reversed. Over the past few decades the exploitation of thickened oil has grown to represent a major part of the petroleum industry. Two really gigantic deposits in Canada and Venezuela account for half the world's supply and together rival the magnitude of the world's total conventional oil accumulations. Despite lying on different continents, both deposits share similar geologic histories. Extraction of this oil has become routine, although environmentally controversial.

Gas hydrates, on the other hand, lie well outside our common experience and may seem like the stuff of science fiction. The very existence of flammable ice on Earth was unknown until the 1960s. The magnitude of hydrate deposits is currently uncertain but clearly huge, with most of them in the form of marine biogenic deposits. Several extraction technologies have been proposed or tested, but it remains to be seen which, if any, will succeed commercially. There is also concern that exploitation could destabilize gas hydrate deposits, with potentially dire consequences for the atmosphere.

A Movable Feast: Petroleum Biodegradation

Crude oil is a complex organic solution that contains literally thousands of distinctly different chemical compounds, ranging from simple methane (CH_4) to complex organic polymers. The microbial specialists that feed on oil have distinct dietary preferences that might be compared to those of small children; they prefer to eat their dessert first and save the vegetables for later (or never). The sweetest molecules are those consisting of a small number of carbon atoms linked together in single file, each carbon holding hands with its neighbors (called n-alkanes; Figure 11.1). In this arrangement each carbon within the chain can also bond with two hydrogen atoms, and the carbons on the ends can bond with three. When they do so, the chain is said to be "saturated" with hydrogen atoms. Chemically speaking such compounds are not much different from the tasty saturated fats that we are frequently told to avoid overeating, but do anyway (for example, see Figure 5.7 for the chemical structure of edible fats called triglycerides).

Once microbes have gobbled up the simplest straight-chain hydrocarbons they start in on molecules in which the carbons form circular 6-carbon daisy chains. These carbon rings themselves link together as building blocks to make larger molecules such as steranes, all of which contain three 6-carbon rings and one 5-carbon ring (the structures of cholesterol, testosterone, and other steroids all follow the same basic pattern). Such compounds may still be saturated with hydrogen, or alternatively the carbon atoms may share most of their electrons with each other instead. Compounds that fall in the latter category are described as aromatic, because they often have a strong odor. Aromatic compounds represent the equivalent of vegetables on the microbial menu; they are not preferred but are still digestible to a certain extent. Many aromatic compounds are toxic to humans, and some are carcinogens.

Microorganisms can consume most light, saturated hydrocarbons and some aromatics, but at some point they push back from the table, having decided that enough is enough. Left untouched on the plate are larger, more complex compounds called resins and asphaltenes. Because they contain nitrogen, sulfur, and oxygen, technically these

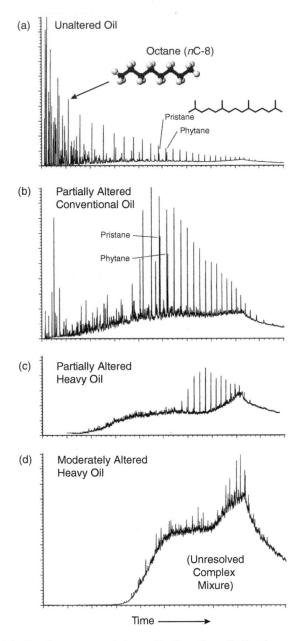

Figure 11.1. Gas chromatograph traces for Canadian oils that have experienced increasing degrees of alteration.

Each peak represents a discrete organic compound. The horizontal axis measures the time required for different compounds to pass through the

compounds are not even hydrocarbons. Their molecular weights reach into the thousands, and asphaltene can be thought of as small bits of kerogen that have been carried along in solution. A variety of different names are commonly used to describe the biodegraded petroleum that is left behind, including tar, asphalt, and pitch, but none of these have really precise definitions. All are encompassed in the more rigorous term "bitumen," which petroleum geochemists use to indicate any material that can be taken into solution by organic solvents.

Most conventional crude oils are less dense than water; that is the reason they migrate upward in the subsurface. Refined products can be lighter still; for example, gasoline has a density only about 75 percent of that of water. Because microbes preferentially remove the smaller, lighter molecules from biodegraded oil, the material they leave behind becomes proportionately denser. Oil that is more dense than normal crude but still less dense than water is therefore termed "heavy," a term that can be loosely applied to biodegraded oils in general. Severely biodegraded oils that have become denser than water are described as "extra heavy." Increased oil density is typically accompanied by increased sulfur content as well, because sulfur resides chiefly in the resin and asphaltene "leftovers."

Oil viscosity, defined as resistance to flow, also tends to increase in degraded, high-density oils and directly influences the techniques required for extraction. Some classifications therefore use viscosity as the principal means of distinguishing conventional oils from unconventional oils. This approach inevitably leads to

Figure 11.1. (*cont.*)

chromatographic column; low molecular weight compounds generally appear first, and higher molecular weight compounds appear later. (a) The dominant peaks in unaltered oil are *n*-alkanes (straight-chain hydrocarbons), with low molecular weight compounds dominant. Sample structure is shown for octane, which has 8 carbon atoms arranged in a chain. A simplified structure is also shown for pristane, which consists of a carbon chain with regular branches. (b) Low molecular weight *n*-alkanes have been removed by biodegradation. (c) A more degraded oil in which only higher molecular weight *n*-alkanes remain. (d) Most of the originally identifiable compounds in the unaltered oil have been removed, leaving behind a complex unresolved mixture that consists of thousands of compounds, which are not separately identifiable and which are resistant to biodegradation (modified from Creaney et al., 1994).

confusion, though, because it lumps together heavy, biodegraded oil with oils that contain a high percentage of wax. The waxes derive from leaf coatings or cell wall membranes in the original organic matter sources and are unrelated to biodegradation. Waxy oil is capable of flowing from a reservoir into a well bore at warm subsurface temperatures, but it thickens or even solidifies into a lightweight substance resembling shoe polish at the cooler temperatures found at the surface. Waxy oils can be extremely viscous, but they have *low* densities. Oil viscosity is valuable in assessing recovery factors, but an oil's density is a better measure of its geologic origins.

Biodegradation by aerobic bacteria requires oxygen, which is primarily dissolved in freshwater. Aerobic biodegradation is therefore confined to depths where freshwater derived from rain and snow can percolate downward. This oxygen requirement does not apply to anaerobic biodegradation by methanogens. Although they operate more slowly than aerobic bacteria, in principal they can biodegrade oil at greater depths. Temperature is also a factor, however. Like the fairy-tale heroine Goldilocks, these microbes also prefer their meal not too cold, and not too hot. Something in the range of 20°C–30°C appears to be "just right," and most microbes lose their appetite entirely at temperatures approaching 80°C. In practice, most heavy and extra heavy oils are found relatively near the surface, and nearly all occur at depths of 2–3 km or less.

Rivers of Oil I: Athabasca

Jasper National Park in the Canadian Rockies ranks among the most ruggedly scenic tourist destinations in the world, drawing about 2 million visitors annually (a number about equal to the population of Vancouver; Figure 11.2). One of its chief attractions is the Columbia ice field, which provides a small glimpse of the dramatically more frigid conditions that prevailed in the area only about 15,000 years ago. The Columbia glacier flows from this ice field, and its meltwaters give rise to the Athabasca River. The river flows northeast and then north through a flatter and more monotonous landscape, toward the forested subarctic lowlands of northeastern Alberta.

As it winds its way north the Athabasca River also erodes downward into its bed, cutting through the deposits of older rivers that followed a similar course about 120 million years earlier. Sandstone

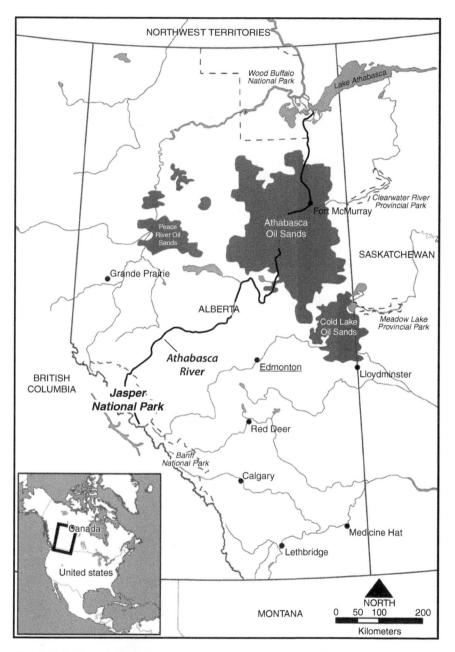

Figure 11.2. Map of Canadian oil sands, showing the course of the Athabasca River (NormanEinstein, Wiki Commons).

outcrops exposed along the Athabasca were reported by 18th century explorers to be impregnated with a thick, black, heavy oil; additional exploration showed that these deposits spread across enormous areas of Alberta with a minor fringe extending into western Saskatchewan. Collectively they represent the single largest in-place oil accumulation in the world, estimated to contain about 1.7 *trillion* barrels of heavy to extra-heavy oil. Although the quantity of this oil is mind-boggling, its quality is poor. Its density varies but generally lies close to that of water. Its viscosity changes with the seasons; it may flow like molasses during the summer, but during the cold northern winters it more closely resembles peanut butter. It is also among the sourest of crudes, with sulfur content commonly near 5 percent.

Why does Canada have so much heavy and extra-heavy oil? A clue arises from the ancient river deposits that are now being eroded by the modern Athabasca. Unlike their modern counterparts, Cretaceous rivers flowed into a shallow sea that encroached across much of central North America, depositing deltas that were partly reworked by waves. The continued rise of this sea eventually buried the river deposits beneath thick layers of marine mud, providing a relatively impermeable seal. The sandstone and mudstone together then continued to be buried beneath additional rock layers for another 60 million years or so, as the ancestral Canadian Rockies pushed eastward. The weight of the mountains caused what are now the plains of Alberta to flex downward, creating a wedge-shaped "hole" within which sediment could accumulate. The resultant sedimentary basin is thickest immediately adjacent to the mountains and tapers continuously to the east (Figure 11.3).

The Western Canada Sedimentary Basin acted as a giant natural collection system for oil generated from multiple source rock intervals distributed across a large geographic area. The efficiency of this gathering system was greatly enhanced by a regional erosion surface, called an unconformity, that formed just before the Cretaceous rivers spread sand across the basin. The unconformity truncates underlying oil source rock intervals at a slight angle. As the source rocks were buried deeper they began to generate oil, which ultimately flowed across the unconformity into the relatively permeable Cretaceous sandstone above. By this time the sandstone layers had themselves been tilted to the southwest, providing natural conduits for buoyant oil to flow northeast. The lengths of these migration pathways were truly impressive, estimated to reach as much as 600 km, resulting in

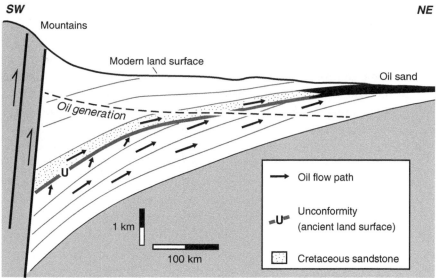

SW

Mountains

Modern land surface

Oil sand

Oil generation

U

→ Oil flow path

⊸U⊸ Unconformity
(ancient land surface)

1 km

100 km

▭ Cretaceous sandstone

NE

Figure 11.3. Schematic cross section of the Western Canada Sedimentary Basin from Jasper National Park to Fort McMurray.

the accumulation of staggering quantities of oil at the basin's thin northeastern edge.

The oil reaching the edge of the basin encountered microbes that lived within the shallow subsurface. Biodegradation greatly increased oil viscosity and increased its density, neutralizing buoyancy-driven oil flow. Evaporation of volatiles and dissolution of water-soluble compounds further contributed to its sedentary nature. Stoppage of oil flow near the surface caused a traffic jam of sorts for new oil traveling up from deeper parts of the basin, arresting it in place until it too became degraded. The resultant deposits today underlie an area of more than 140,000 km², roughly equal in size to the state of New York, and extend from the surface down to depths of up to 1 km.

The time window for generation closed shortly after the extinction of the dinosaurs. At about 60 million years ago new deposits ceased to accumulate in the basin and the old ones began to erode, as hundreds to thousands of meters of rock were removed from the land surface. Stripping of those shallower rock layers moved underlying oil source rocks closer to the surface, eventually cooling them to temperatures at which they could no longer generate new oil. According to calculations by geologist Debra Higley and colleagues at the U.S.

Geological Survey, generation had ceased entirely by about 40 million years ago.

The exact amount of oil that has been generated, and the detailed contributions of different candidate source rocks, are matters of geologic controversy. Two former Mobil Oil geologists, Stephen Moshier and Douglas Waples, estimated in 1985 that at least 6 trillion barrels of oil must have originally been generated, in order to account for the oil that remains today. Most of that original total has been lost, in part because the Albertan plains lack the giant traps found in places like the Persian Gulf, and in part because the Alberta source rocks have long ceased to generate new oil to replace that lost to biodegradation and seepage. The preserved deposits therefore represent a small remnant of what was once a much larger volume of oil; Albertans can only dream about "what might have been."

Rivers of Oil II: Orinoco

Canada shares little in common with Venezuela; in fact the two countries could hardly appear more different in geography, climate, population, language, culture, or cuisine. They both possess uncommon endowments of oil, however, and in some ways each might be thought of as the "Saudi Arabia" of its respective continent. More specifically, Canada and Venezuela hold the world's two largest accumulations of heavy to extra-heavy oil.

The geologic history of heavy to extra-heavy oil deposits near the Orinoco River in Venezuela in many ways resembles that of the Athabasca deposits. Both formed within wedge-shaped basins by long-distance migration of oil from deeply buried source rocks. In both cases oil became biodegraded near the surface. Both oil accumulations are contained partly within the deposits of ancient rivers that followed paths similar to their eponymous modern equivalents. Estimated in-place heavy to extra-heavy oil reserves are roughly comparable in both countries, with Venezuela estimated to hold something around 1.3 trillion barrels. Oil properties are also fairly similar on the whole, although they vary within each area. The Venezuelan biodegraded oils typically have densities slightly greater than that of water, whereas the Canadian oils are commonly somewhat lighter. Countering this difference, the Venezuelan oil may actually flow more easily, because average annual surface temperatures are on the order of 20°C higher in Venezuela.

The geology of Venezuelan heavy and extra-heavy oil also differs from that of western Canada in some significant ways. Rather than forming in the interior of a major tectonic plate, the Orinoco deposits formed near the complex boundary between the Caribbean and South America Plates. Normally in such situations the oceanic crust (the Caribbean) would be expected to subduct beneath the continental crust (South America), because of its greater density and lesser thickness. Northern Venezuela is a rare exception, however, where the continental crust has been pushed down beneath the adjacent oceanic crust. The Eastern Venezuela sedimentary basin is therefore thickest adjacent to the Caribbean and thins inland, to the south. It is only about one-third the width of its Canadian cousin, a difference that puts it at a disadvantage in terms of source rock area available to drain. However, unlike in western Canada, oil generation is still active in Venezuela; the accumulation of Orinoco oil began only about 15 million years and continues today. The Eastern Venezuela basin thus represents a "live" petroleum system rather than a preserved relict and is especially prolific for its size.

The U.S. Geological Survey has estimated that about 500 billion barrels might ultimately be recovered from Orinoco deposits, versus the official Canadian estimate of about 175 billion barrels made by the Alberta provincial government. The relative significance of these numbers is unclear, however, since they were derived differently. Currently the Canadian deposits produce more oil.

Making the Most of It: Heavy-Oil Extraction and Upgrading

Canadian and Venezuelan oil sands both lie at burial depths of 1 km or less. Unlike Venezuela's, the Canadian deposits also extend upward to the surface, and the simplest means of extraction therefore is to mine the oil sand from large open pits. As is true with coal mining, surface mining of Canadian oil sand carries a practical depth limit, theoretically about 75 m, although most mines are shallower than this. An estimated 20 percent of the total recoverable resource is accessible through surface mining. Once mined, a combination of hot water and organic solvents are used to separate the oil from the sand. Recovery efficiencies for surface mining are high, on the order of 75 percent or more. The process is labor- and capital-intensive, however, and results in permanent alteration of a landscape that is underlain mostly by muskeg, a kind of water-logged peat deposit. Reclamation of mine sites and associated tailings ponds can help hide the traces of mining but cannot replace peat lands that took thousands of years to develop.

Deposits not within reach of surface mining must be accessed by drilling, combined with some method of compelling the reluctant oil to flow out of the reservoir. Many different technologies have been devised to achieve this end, resulting in a proliferation of different processes known chiefly by their acronyms. For example, CHOPS stands for "cold heavy oil production with sand." This technique involves using strong pumps to extract oil from the ground, along with some of the sand within which the oil resides. This only works for some of the less viscous deposits, however, and even then recovery rates are generally less than 10 percent. Higher recovery rates require heating the oil before extraction. In CSS (cyclic steam injection, colloquially known as "huff and puff"), superheated steam is injected into the ground, and then the rocks are allowed to heat-soak for a period of hours or days. Finally, the heated, less viscous oil is pumped to the surface through the same well that received the steam. The cycle is repeated until the amount of oil extracted declines below an economic limit. Recovery efficiency using CSS may rise to 20–30 percent. Locally recovery efficiencies as high 60–70 percent can be obtained using a more recent technique called SAGD (pronounced "sag-dee," for steam assisted gravity drainage). In this technique two parallel horizontal wells are drilled approximately 6–7 meters apart, one above the other. Steam is injected into the upper well, and oil plus water are produced from the lower well.

Gasoline and jet fuel are the glamour products of the fossil fuel world; they perform the most prodigious feats and command the highest prices. Unfortunately they have been largely eradicated from heavy and extra-heavy oil by ravenous microbes. What initially emerges from the ground instead resembles the dirty, smelly stuff used in road paving, roof shingles, and other applications where we want the cheapest sealant that we can get, no questions asked. In addition to high sulfur content it has an affinity for heavy metal, including nickel, vanadium, iron, lead, chromium, mercury, arsenic, and selenium. Much work lies ahead to upgrade this smelly, toxic goo into salable synthetic crude oil (SCO), from which valuable transportation fuels can be refined.

The process begins with separating sand and water from the oil, followed by moderate heating and distillation to recover any light hydrocarbon compounds that the microbes may have missed. The asphaltic remainder is then heated at higher temperatures (>450°C) to break apart thermally its large molecules into smaller ones. This process is known as "coking" because it produces a carbon-rich residuum (coke) as a by-product. It is chemically similar to the retort process used to break down oil shale kerogen (see Chapter 8). An alternative approach is "hydrocracking," in which water and a catalyst are used to help crack larger molecules apart. Subsequent hydrotreating further upgrades the SCO, which can then be refined into specific products in a manner similar to the processing of conventional crude. The details of the overall processing sequence vary, but the goals are always to increase the relative proportion of light hydrocarbon compounds (transportation fuel) contained in the final product.

Surface mining of oil sand has long been controversial because of its heavy usage of water; by some estimates 2 to 4.5 barrels of water is required to make a barrel of synthetic crude. Most of this water currently is obtained from the Canadian Rocky Mountains, by way of rivers such as the Athabasca, or from groundwater. In situ recovery methods such as SAGD consume less water but more energy, because of the need to superheat steam. Assuming that fossil fuels provide this energy, it adds significantly to the greenhouse gas emissions associated with oil sands. Additional CO_2 is released during the upgrading of heavy or extra-heavy oil. It has been estimated that on average, fuels derived from heavy and extra-heavy oil ultimately are responsible for up to 20 percent more CO_2 than those derived from conventional light oil.

Frigid Fuel: Gas Hydrates

Under the right conditions of temperature and pressure, methane and other gases bubbling up from the warm depths may become imprisoned by three-dimensional cagelike structures of ice, which confine gas molecules within. The generic term for this type of compound substance is "clathrate," derived from a Latin word meaning lattice, bars, or railing. Clathrates in which the cage is built from water molecules are known more specifically as clathrate hydrates, and those that contain methane are commonly referred to simply as methane hydrate (or more generally as gas hydrate). The tight quarters imposed by the hydrate structure pack methane molecules closely together, increasing their density by a factor of 164 compared to methane gas at surface atmospheric pressure.

Gas hydrates form when gas and water meet under conditions of low temperature and elevated pressure. On land, the pressure is applied by the weight of pore water contained in overlying

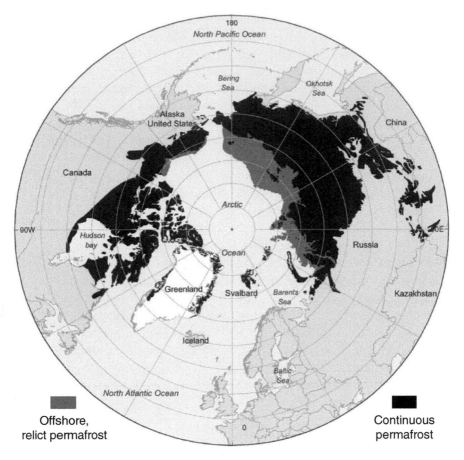

Figure 11.4. Permafrost map of the Northern Hemisphere.

Note that methane hydrate deposits may exist beneath broad areas of northern Asia and North America (from Collett et al., 2001).

rocks and sediment; it increases predictably with burial depth. The cool temperatures required are only attainable in high arctic locales, where average annual surface temperatures fall below 0°C (Figure 11.4). Such areas are typically capped by permafrost, consisting of water-saturated sediment and rocks that are permanently frozen. Methane hydrate can form below the permafrost at depths as shallow as 200 m at temperatures of about −10°C. Its freezing point increases to about +15°C at depths of 1 km or greater. Its lower bound is defined by the point at which subsurface temperatures begin to exceed this limit.

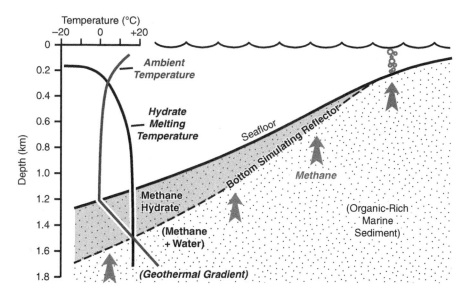

Figure 11.5. Schematic illustration of the stability zone for methane hydrate beneath the seafloor.

In theory, methane hydrate is stable anywhere the black curve crosses to the right of the gray curve. In practice it is found only within the sediment, because of low gas saturation in seawater.

Conditions for hydrate formation are more widespread on the seafloor adjacent to the continents (Figure 11.5). The deep waters of the ocean are surprisingly cold, because of the conveyorlike circulation of ocean waters around the planet (see Figure 7.7). The ocean's bottom waters largely originate in the northern Atlantic near Greenland, an area known for icebergs such as the one that sank the RMS *Titanic* in 1912. These dense, near-freezing waters sink to the bottom in the North Atlantic, spread southward, and then flow eastward toward the Indian and Pacific Oceans. As they travel they gradually warm up, then return to the Atlantic as surface flow about one thousand years after they start their journey.

The deep ocean also supplies the pressure required for hydrate formation. In principal gas hydrate can be stable at water depths as shallow as a few hundred meters, conjuring up images of clouds of fluffy white hydrate "snow." Unlike normal snow, hydrate would float upward, because it is less dense than seawater. Such inverted snowstorms do not generally occur, though, because the concentration of natural gas in seawater is too low. Gas can reach higher

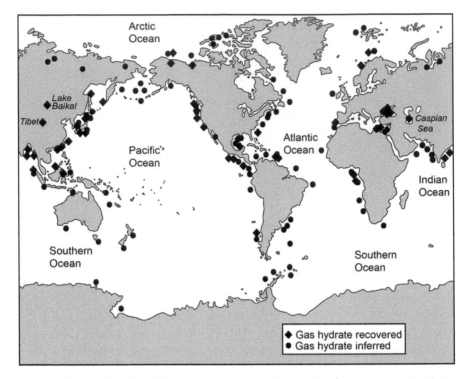

Figure 11.6. Localities where gas hydrates have either been recovered during drilling or inferred to exist from seismic surveys (map compiled by Carolyn Ruppel and Diane Noserale of the U.S. Geological Survey).

concentration within the sediment below the seafloor, where it displaces some of the water that initially fills pore spaces. Methane is emitted from methanogens feeding off organic-rich muds below and then rises toward the seafloor through natural fractures or interconnected networks of pores. Solid deposits form at the point where this upward-migrating natural gas encounters sediment at a temperature cool enough for methane hydrate to be stable. This temperature generally falls between 0°C and +15°C, depending on how much pressure is exerted by overlying seawater.

As noted in Chapter 7, organic-rich marine mud deposits cluster near the edges of the continents, because of a favorable combination of high nutrient supply and diminished bottom-water oxygenation. Gas hydrate deposits follow the same general pattern and occur nearly ubiquitously adjacent to most continental margins (Figure 11.6). Although the majority of such deposits contain biogenic methane, natural gas generated by geothermal heating within sedimentary basins is also commonly entrapped as hydrate.

The Size of the Prize(?)

Estimates of the total amount of natural gas locked up in hydrate vary widely, in part because detailed drilling data are sparse. The general location of marine hydrate deposits is often easy to deduce, however, as a result of the strong and relatively abrupt density contrast that occurs between the bottom of the hydrate-bearing zone and underlying sediment that is saturated with free natural gas. Sound waves projected below the seafloor during seismic surveys bounce off this transition, providing a clear image of the bottom of the hydrate-bearing zone. The position of this surface is governed by the generally uniform rate at which temperatures increase with burial depth. As a result, the shape of the surface tends to replicate the shape of the overlying seafloor. This surface is therefore termed a "bottom-simulating reflector" (Figures 11.5 and 11.7), which allows the potential thickness of hydrate-bearing sediment beneath the seafloor to be easily mapped.

Seismic investigations generally do not measure the actual hydrate saturation within this interval, however; drilling is therefore needed to address resource potential fully. This potential varies widely, depending on reservoir porosity and permeabilty and on the degree of saturation of pore spaces with hydrate. Nonetheless, conservative estimates place the total global magnitude of gas hydrate in range of 5,000 gigatons of carbon, which if accurate would equal or exceed the magnitude of all other known fossil fuels combined. It must be noted, however, that the hydrate estimate represents in-place resources, whereas fossil fuel estimates normally target recoverable reserves.

If gas hydrate deposits are so gigantic, why aren't we using them already? Primarily because no commercially practical means of recovering them has yet been demonstrated. A number of techniques have been proposed and even tested, including heating of the sediments, pumping to decrease pressure (which destablizes hydrates), and injecting methanol or other chemical antifreeze to break open the hydrate cages. One of the more intriguing proposals is to inject CO_2, in a sort of prisoner exchange that simultaneously liberates natural gas and disposes of "used" carbon. Even if commercial extraction does become feasible, it will likely focus only on the richest deposits, leaving behind the leaner deposits that constitute the majority of global

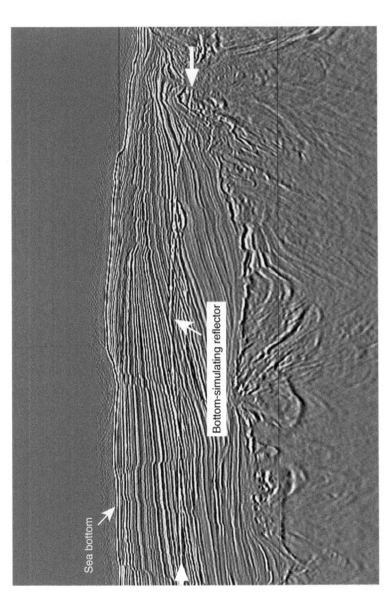

Sea bottom

Bottom-simulating reflector

Figure 11.7. Offshore seismic record from the Kumano Basin offshore of southern Japan, spanning approximately 8.3 km horizontal distance. The gray in the upper quarter of the image is seawater; continuous reflections below the sea bottom (dark and light lines) correspond to tilted sedimentary layers. Note that the bottom-simulating reflector (BSR) parallels the sea bottom and cuts across many of the reflections from sedimentary layers (image courtesy of Harold Tobin and the Japan Agency for Marine-Earth Science and Technology).

hydrate volume. Hydrate associated with permafrost appears to be the most economically accessible and has been the target of pilot projects in Alaska and Siberia. Deep ocean deposits are by their nature more expensive to pursue but may nonetheless be of interest in regions lacking other natural gas resources. Japan, for example, has been investigating hydrate extraction near the Nankai Trough, an offshore area south of Honshu that is much better known for earthquakes than fossil fuels.

Finally, it has been suggested that catastrophic release of methane from hydrate could effectively bring about the end of the world as we know it, which would certainly qualify as "a bad thing"! The idea is that extraction efforts might inadvertently result in a large release of methane to the atmosphere, for example, by triggering submarine landslides. Because the short-term global warming potential of methane is much higher than that of CO_2, a sudden massive release would result in dramatic global warming. Higher temperatures would in turn further destabilize other methane hydrate deposits, potentially leading to a runaway chain reaction. Geologist James Kennett labeled this the "clathrate gun hypothesis."

While such a doomsday scenario might seem bizarre, it appears that something like this may have actually happened about 55 million years ago, triggering a period of abnormally warm climate known as the Paleocene-Eocene Thermal Maximum (PETM; see discussion in Chapter 2). Interestingly, land-dwelling mammals increased greatly in diversity after this event, apparently making the most of the unexpectedly warm weather. Good news for mammals turned out to be bad news for bottom-dwelling sea creatures, though, many species of which perished. The causes of the PETM continue to be debated in scientific circles, but it clearly involved the unusually rapid release of organic carbon into the atmosphere.

Fortunately it appears that a methane release of the magnitude implied for the PETM would likely require millenia rather than decades. Geologist David Archer of the University of Chicago noted in a 2007 study that most of the mass of existing hydrate lies buried deeply within sedimentary layers beneath the seafloor and that sediment acts as an effective insulator. It would therefore take thousands of years for heat to penetrate from the seafloor downward to these deposits. Furthermore, the thousand-year circulation of the ocean means that atmospheric heat would take a long time to warm bottom waters enough to destablize hydrates. The history of the PETM itself shows that the onset of higher temperatures was relatively gradual, when measured in human terms. Rather than a catastrophic explosion of methane from hydrate, we are more likely to face a gradually growing addition to the greenhouse gases already being released from the combustion of other fossil fuels.

Conclusions

"Petrified" oil and gas start out as freely migrating fluids, and therefore they initially obey the same basic principles that apply to conventional oil and gas. They must be generated from a source rock or sediment, migrate upward toward the surface, and become concentrated in a trap. Like conventional oil and gas deposits, the richest and most commercially viable oil sand and gas hydrate deposits can be expected where a number of different geologic factors fortuitously coincide. The primary distinguishing feature of biodegraded oil and gas hydrate is that they become immobile after undergoing migration; they therefore aid in their own imprisonment.

The in-place magnitude of oil sand and gas hydrate deposits is clearly enormous. The total global resource of heavy and extra-heavy oil has been estimated at somewhere near 6 trillion barrels, which is about twice the world's original endowment of conventional oil as estimated by the U.S. Geological Survey. Such large numbers are potentially misleading, however, because the majority of this resource exists either within poor-quality reservoirs or at low levels of reservoir saturation. Currently only about 10 percent of the oil in Canadian oil sand is considered to be economically recoverable, and this may prove to be a good rule of thumb elsewhere.

Commercial extraction of gas hydrate has not yet commenced, so the magnitude of the recoverable resource remains speculative. When (or if) commercial extraction does occur it will undoubtedly focus on the richest deposits. The most economically attractive deposits will likely be those associated with permafrost, because drilling and production onshore are far cheaper than in the deep ocean. Onshore hydrate deposits only account for about 1 percent of the total global resource, however, and high-quality deposits represent a small fraction of that 1 percent.

The reanimation of petrified fossil fuels will likely require larger amounts of input energy than would be needed for the extraction of conventional oil and gas. Heat input is required to separate heavy and extra-heavy oil from its host reservoir (or from mined sand) and for upgrading to synthetic crude oil. Extraction of natural gas from hydrate also will require substantial energy input to destabilize these solid deposits.

Extraction of heavy and extra-heavy oil inevitably disrupts the landscape, particularly if mined at the surface, and consumes larger volumes of water than conventional oil and gas production. Oil sands also emit more greenhouse gases over their life cycle. In contrast to these now-familiar environmental problems, the full implications of gas hydrate extraction have yet to be demonstrated. The provocative clathrate gun hypothesis may be far-fetched, but it nonetheless seems prudent to treat methane hydrate as a loaded weapon.

For More Information

Aitken, C. M., Jones, D. M. and Larter, S. R., 2004, Anaerobic hydrocarbon bio-degradation in deep subsurface oil reservoirs: *Nature* 431, 291–294.

Archer, D., 2007, Methane hydrate stability and anthropogenic climate change: *Biogeosciences*, v. 4, p. 521–544.

Collett, T. S., Lee, M. W., Agena, W. F., Miller, J. J., Lewis, K. A., Zyrianova, M. V., Boswell, R., and Inks, Tanya, 2011, Permafrost-associated natural gas hydrate occurrences on the Alaska North Slope: *Marine and Petroleum Geology*, v. 28, p. 279–294.

Creaney, S., Allan, J., Cole, K. S., Fowler, M. G., Brooks, P. W., Osadetz, K. G., Macqueen, R. W., Snowdon, L. R., and Riediger, C. L., 1994, Petroleum generation and migration in the Western Canada sedimentary basin, *in* Geological Atlas of the Western Canada Sedimentary Basin, G. D. Mossop and I. Shetsen (comp.), Canadian Society of Petroleum Geologists and Alberta Research Council (http://www.ags.gov.ab.ca/publications/wcsb_atlas/atlas.html).

Dusseault, M. B., 2001, Comparing Venezuelan and Canadian heavy oil and tar sands: Canadian International Petroleum Conference, Paper 2001–061, 20 p. (www.energy.gov.ab.ca/oilsands/pdfs/rpt_chops_app3.pdf)

Gosselin, P., Hrudey, S. E., Naeth, M. A., Plourde, A., Therrien, R., Van Der Kraak, G., and Zu, Z., 2010, Environmental and Health Impacts of Canada's Oil Sands Industry: Ottawa, Canada, *Royal Society of Canada*, 414 p.

Hein, F. J., 2000, Historical Overview of the Fort McMurray Area and the Oil Sands Industry in Northeast Alberta: *Alberta Energy and Utilities Board, Earth Sciences Report* 2000–05, 32 p.

Hester, K. C., and Brewer, P. G., 2009, Clathrate hydrates in nature: *Annual Reviews of Marine Science*, v. 1, p. 303–327.

Higley, D. K., Lewan, M. D., Roberts, L. N. R., and Henry, M., 2009, Timing and petroleum sources for the Lower Cretaceous Mannville Group oil sands of northern Alberta based on 4-D modeling: *American Association of Petroleum Geologists Bulletin*, v. 93, p. 203–230.

Humphries, M., 2008, North American Oil Sands: History of Development, Prospects for the Future: *Congressional Research Service Report* RL 34258, 27 p.

James, K. H., 2000, The Venezuelan hydrocarbon habitat, Part 1: Tectonics, structure paleogeography, and source rocks: *Journal of Petroleum Geology*, v. 23, p. 5–53.

James, K. H., 2000, The Venezuelan hydrocarbon habitat, Part 2: Hydrocarbon occurrences and generated-accumulated volumes: *Journal of Petroleum Geology*, v. 23, p. 133–164.

Kennett, J. P., Cannariato, G., Hendy, I. L., and Behl, R. J., 2003, Methane Hydrates in Quaternary Climate Change: The Clathrate Gun Hypothesis: Washington DC, American Geophysical Union, 216 p.

Lattanzio, R. K., 2013, Canadian Oil Sands: Life-Cycle Assessments of Greenhouse Gas Emissions: *Congressional Research Service Report* R42537, 31 p.

Moshier, S. O., and Waples, D. W., 1985, Quantitative evaluation of Lower Cretaceous Mannville Group as source rock for Alberta's oil sands: *American Association of Petroleum Geologists Bulletin*, v. 69, p. 161–172.

Ruppel, C., 2011, Methane Hydrates and the Future of Natural Gas: Supplementary Paper #4, The Future of Natural Gas, MIT Energy Initiative study, 25 p.

Schenk, C. J., Cook, T. A., Charpentier, R. R., Pollastro, R. M., Klett, T. R., Tennyson, M. E., Kirschbaum, M. A., Brownfield, M. E., and Pitman, J. K., 2009, An Estimate of Recoverable Heavy Oil Resources of the Orinoco Oil Belt, Venezuela: *U.S. Geological Survey Fact Sheet* 2009–3028, 4 p.

Suess, Erwin, Bohrmann, G., Geinert, J., and Lausch, E., 1999, Flammable ice: *Scientific American*, v. 281, p. 76–83.

Summa, L. L., Goodman, E. D., Richardson, M., Norton, I. O., and Green, A. R., 2003, Hydrocarbon systems of Northeastern Venezuela: Plate through molecular scale-analysis of the genesis and evolution of the Eastern Venezuela Basin: *Marine and Petroleum Geology*, v. 20, p. 323–349.

Vigrass, L. W., 1968, The geology of Canadian heavy oil sands: *American Association of Petroleum Geologists Bulletin*, v. 52, p. 1984–1999.

12

Water, Water, Everywhere

When the well's dry, we know the worth of water.

Benjamin Franklin (1706–1790), Poor Richard's Almanac, *1746*

Water, water, everywhere,
And all the boards did shrink;
Water, water, everywhere,
Nor any drop to drink.

Samuel Taylor Coleridge, "The Rime of the Ancient Mariner," 1798

Water is without doubt the most vital natural resource on Earth. Humans are mostly made of water, and without continual replenishment we cannot expect to live more than a few days. Even if we could, the plants that we depend upon for sustenance would quickly wither and die without water. Fortunately water is abundant on Earth, so much so that we tend to overlook its importance. The other planets in the solar system are not so fortunate, and that probably accounts for the apparent absence of life there. There is a chance that Mars did harbor life in the distant geologic past, and much of our exploration of that planet has focused on its history of water. Numerous canyons, dry river channels, and layered sedimentary deposits attest to the importance of water in originally shaping the Martian landscape. It has since become a hyperarid planet, however, where water appears to be confined to limited areas of polar permafrost.

The geological record of the Earth's first few hundred million years has almost entirely vanished, but what little evidence remains suggests that liquid water was already present at the Earth's surface as early as 4.4 billion years ago. Fortunately for us the Earth's overall water supply appears to be stable. By far the largest water resource is the ocean, which presently covers 71 percent of the planet's surface to a mean depth of nearly 4,000 m. As noted in Chapter 2, over the past 541 million years the depth of the ocean has remained surprisingly consistent, fluctuating only a few hundred meters at most.

There is really no chance humans could ever exhaust the planet's water supply, but possession of inexhaustible quantities of seawater is not cause for complacency. Seawater is grossly unfit for human consumption, and it is also poisonous to most agricultural crops. Freshwater constitutes only about 2.5 percent of the Earth's total near-surface supply, with about two-thirds of that amount currently locked away in inaccessible glacial ice (Figure 12.1). Even if global warming melts some of the ice, the released water will mostly end up in the sea. Less than 1 percent of the Earth's near-surface water supply resides in freshwater aquifers, rivers, streams, lakes, and other bodies of water that are potentially suitable for sustaining life on the continents.

Our challenge then is to support a rapidly growing world population using the relatively small amount of water that is fresh, liquid, and accessible. This is by no means an easy challenge, and so far we have not been doing a very good job of meeting it. The United Nations Environmental Program estimated that by 2025 two-thirds of the world's population will live in "water-stressed" areas. Even in areas that possess potentially adequate supplies, water pollution by agricultural, industrial, and human wastes often compromises their quality. An estimated 20 percent of the global population already lacks access to clean water, and waterborne diseases are currently the second-leading cause of death worldwide.

Food production currently uses the greatest amount of freshwater, accounting for about 75 percent of net global use according to the United Nations Environmental Program. Most of this water is for irrigation in areas that either lack sufficient rainfall to grow crops or else have soils that don't retain water very well. Energy production ranks a surprisingly close second in gross water usage, primarily because water is used to carry away the waste heat generated by thermoelectric power plants (both fossil fuel– and nuclear-powered). In the United States, thermoelectric cooling accounts for about a third of total freshwater withdrawals. Fortunately, most of this cooling water is returned to its source afterward, warmer but otherwise none the worse for wear. The principal losers are downstream flora and fauna that may not appreciate the "improved" temperatures. The *net* loss of freshwater associated with thermoelectric cooling is relatively small, around 3 percent of total U.S. water consumption.

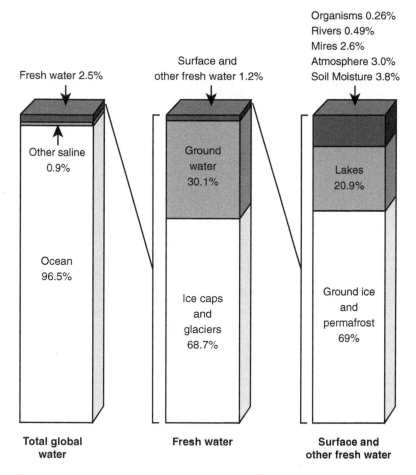

Figure 12.1. Distribution of water on Earth (modified from U.S. Geological Survey).

Virtually every other system for producing or utilizing energy is also linked inextricably to water. Unsurprisingly, biofuels are by far the biggest users of freshwater relative to the amount of energy they provide. Fossil fuel extraction and refining consume lesser, but still substantial amounts of water. Primary water use is only part of the problem, however; contamination of freshwater supplies can effectively render them unusable. Furthermore, energy must be expended to raise water from a well, pump it to its final destination, purify it, or heat it for a shower. Energy can even be used to extract freshwater from the sea by desalination, an effective but costly solution to the paradox faced by Coleridge's "Ancient Mariner."

What's So Special about Water?

Water is the universal solvent, possessed of almost magical properties that are essential to life as we know it. Water also shapes the landscape around us, through a wide array of processes ranging from freeze-thaw to river erosion. It even helps to build the continents, by lowering the temperature at which the relatively dense rocks of the Earth's mantle and seafloor begin to melt. Partial melting creates magma that rises upward and recrystallizes as continental rocks. The volcanoes of the Andes Mountains, for example, formed as a result of subduction and partial melting of water-saturated seafloor beneath the western edge of South America.

Every high school chemistry student knows that water is a chemical compound consisting of two hydrogen atoms and one oxygen atom, with the chemical formula H_2O. These elements are by themselves exceedingly common; hydrogen is the most abundant element in the solar system, and oxygen is the most abundant element in the Earth's crust. As a compound they are considerably more rare, however, and water has some rather unique properties that are related to the *way* in which hydrogen and oxygen combine.

Hydrogen and oxygen are bound to each other through sharing of each other's electrons, but it's not an equal partnership. Electrons have a negative electrical charge and are strongly attracted to the twelve positively charged protons in oxygen nucleus. The two hydrogen protons align themselves 104.5° apart, giving the water molecule a crooked appearance (Figure 12.2). This arrangement also gives water two distinct electrical poles: a negative pole associated with the electrons clustered about the oxygen nucleus and a positive pole associated with the two hydrogen protons.

The electrical dipole formed by water molecules is what gives them their unique properties, because the positive pole of one water molecule tends to attract the negative poles of other water molecules. This attraction, known as hydrogen bonding, pulls the water molecules more tightly toward each other. As a simple analogy you might think of a bag filled with magnets, in which the north pole of one magnet is attracted to the south pole of another. Collectively the magnets will tend to clump together as a single, contorted mass, with individual magnets lining up into tightly bound but irregular three-dimensional patterns. The electrical attraction between the positive and negative poles of different water molecules causes a similar effect.

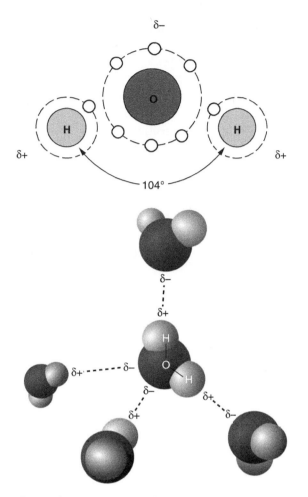

Figure 12.2. Top: Molecular structure of water.

Small white circles are electrons; note that clustering of the electrons near the strongly positive oxygen nucleus results in an electrostatic dipole, with negative charge associated with the oxygen and positive charges associated with the two hydrogens.

Bottom: Schematic three-dimensional rendering of the interaction between adjacent water molecules.

Hydrogen bonds (dashed lines) form between the positive pole of one molecule and the negative pole of another (modified from Wikipedia Commons).

Water has a relatively high boiling point because the mutual attraction of its molecules keeps them in liquid rather than gaseous form. Liquid water can also carry a lot of heat, because the tight hydrogen bonding inhibits vibration between molecules. The polar nature of water also allows it to dissolve other compounds that are themselves polar. Water acts as an excellent solvent for chemical ions, which are atoms with either a positive or a negative electrical charge. The negative pole of water attracts positively charged ions, for example, sodium, and its positive pole attracts negative ions, for example, chlorine. In solid form these two ions join together to make table salt; in dissolved form they make the ocean salty.

About the only substances that don't dissolve in water are nonpolar molecules called "lipids," derived from the Greek word for "fat." Lipids, which also include the major compounds found in oils, are therefore considered "hydrophobic" (from Greek roots that mean "water-fearing"). Dissolution of oils therefore requires assistance from a detergent such as dish soap, which possesses both polar and nonpolar properties.

Water and Energy

The unique physical and chemical qualities of water make it indispensable to most forms of energy procurement. The actual quantity of water needed to acquire a given amount of useful energy varies widely, however. The U.S. Department of Energy attempted to summarize this variability in a 2006 report to Congress, which compared water use by different energy extraction and power generation systems (Figure 12.3). The report expressed its results in terms of gallons of water consumed per million British Thermal Units (BTU) produced, but these same data also can be viewed in terms of barrels of water consumed per barrel of oil (or its energy equivalent) produced.

At the low end, electricity generation via solar photovoltaic and wind power consumes negligible amounts of water, primarily used to clean the equipment. The situation changes, though, if this electricity is used to make hydrogen fuel; electrolysis consumes about one barrel of water to produce hydrogen with the same energy as one barrel of oil. This is more water than needed to produce an actual barrel of conventional oil. Conventional oil extraction does consume water for use in drilling mud and other purposes, but the amounts are relatively small compared to the large amounts of energy obtained.

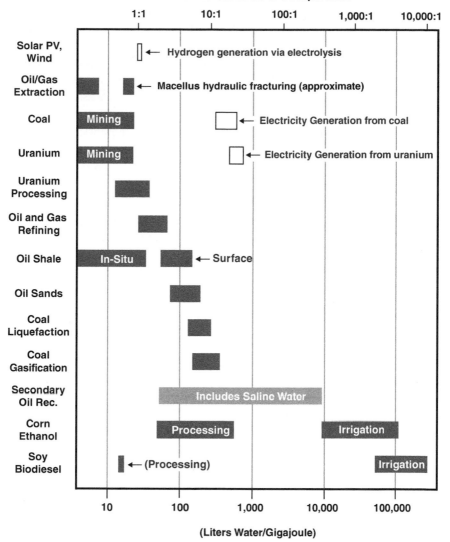

Figure 12.3. Comparison of water consumption associated with various forms of fossil fuel, nuclear, and biofuel energy production.

Note that the horizontal scales are logarithmic; each division represents a tenfold difference in water use (data from Gleick, 1994, and U.S. Department of Energy Report to Congress, 2006).

According to the U.S. Department of Energy, coal and uranium mining also use less than a barrel of water to produce the energy equivalent of a barrel of oil. This water is used in many ways, such as equipment cooling, ore processing and transport, dust suppression, and revegetation of reclaimed mine sites. By far the largest water use associated with coal and uranium occurs when they are used to generate electricity, however, because of the cooling losses associated with thermoelectric power plants.

Coal and uranium mining also carry the potential to contaminate larger volumes of freshwater that are not directly involved in energy production. For example, the sulfur contained in many coals, especially those in the eastern United States, combines readily with water to make sulfuric acid, which can drastically reduce surface water quality if allowed to escape from a mine or its tailings. Sulfur-bearing minerals are commonly associated with uranium deposits and can have a similarly unfortunate impact. Coal and uranium deposits also share an affinity for toxic heavy metals, which can potentially escape from mining operations.

Hydraulic fracturing has become infamous for its thirst for water. Water serves as an inexpensive medium for transmitting the pressures needed to crack open low-permeability reservoirs. Other fluids, such as liquefied natural gas, could be used to do the same job, but they cost more and introduce their own problems (such as explosion hazard). Hydraulically fractured wells drilled in the Marcellus Shale in the eastern United States have been reported to use anywhere from 2 to 9 million gallons per well, which certainly seems like a lot. These wells also return a lot of natural gas, though. According to the U.S. Geological Survey's 2011 assessment of Marcellus Formation gas resources, each well on average is expected to produce about 1 billion cubic feet of natural gas. Assuming that an average well requires 5 million gallons of water for hydraulic fracturing, this works out to slightly less than one barrel of water to obtain the energy equivalent of one barrel of oil. From this perspective hydraulic fracturing doesn't look much worse than conventional oil and gas extraction.

Other forms of oil production are considerably thirstier; for example, the National Energy Board of Canada in 2006 estimated that 2.5 to 4 barrels of water are used to produce a barrel of oil from heavy oil deposits in Alberta. Final refining of these heavy and extra-heavy oils consumes even more water. Production of other synthetic fuels can use more water still; for example, the U.S. Department of Energy reported that coal liquefaction requires between 4.6 and 6.8 barrels of water per barrel of oil.

Secondary oil recovery (see Chapter 9) requires more water than any other form of fossil fuel extraction. This technique works because water adheres naturally to the surfaces of mineral grains in a reservoir but doesn't mix easily with oil. Water flooding therefore can push out large volumes of oil that did not flow out of its own accord because it was "stuck" in pore spaces. Although the quantities injected are prodigious, much of this is water that does not have other uses, such as seawater or water produced from deep saline aquifers. Reinjection of waters produced from deep wells has little or no impact on freshwater supplies, although it could conceivably introduce contamination risks similar to those imputed to hydraulic fracturing. Water flood has been widely used for decades, though, without incurring significant problems.

Biofuels consume by far the largest quantities of freshwater relative to their energy output. For example, a 2009 study by Yi-Wen Chiu and colleagues at the University of Minnesota concluded that anywhere from 5 to 2,138 liters of water was consumed to produce 1 liter of ethanol. They further noted that water appropriation for bioethanol production in the United States increased 246 percent between 2005 and 2008, whereas bioethanol production only increased 133 percent. Water usage for bioethanol falls into two major categories: water used to grow crops and water used to process crops into liquid fuel. Of course, not all crops need to be irrigated; most of the U.S. Corn Belt, for example, relies on natural precipitation. This "free" water falling from the sky explains the lower end of the water consumption estimates made by Chiu and associates. Expansion of biofuels crops into marginal lands with less natural precipitation accounts for the upper end, since such crops would likely require irrigation. Setting irrigation aside, ethanol refining alone consumes as much or more water than conventional petroleum production and refining combined.

Where do we get fresh water? Ultimately it comes from the ocean via evaporation and subsequent precipitation onto the continents, a natural process of solar desalination. The full story is a bit more complicated, however, and requires a more detailed understanding of the travels of water.

Restless Reservoirs: The Hydrologic Cycle

The ancient Greek philosopher Heraclitus famously observed that "you cannot step twice into the same river," meaning that water in a river is in constant motion. At a larger scale you can think of all the world's rivers and streams collectively as a single large reservoir of water, which is continuously flowing into another even larger reservoir, the ocean. If not replenished the world's major rivers would run dry in a matter of months, a fact that can be easily established by comparing length of any given river to the average velocity at which it is flowing. For example, the Amazon River flows at rates in the range of 2.5 to 5.0 km per hour, and is about 6,400 km in length. A log that falls into the headwaters of the Amazon should therefore take two to three months to float down to the Atlantic. If no additional water were available to refill the river channel, it would run completely dry within this time.

Of course, not all of the water in the Amazon has to make the full 6,400 km trip; a log dropped in at the halfway point would spend less time reaching the sea. The lower half of the river also contains a disproportionate amount of the river's water, because the channel becomes gradually wider and deeper going downstream. Some water can also be expected to linger in lake, ponds, and mires encountered along the river course. A convenient way of sweeping all such details under the carpet is to talk only about the *average* amount of time that water resides in the river. Average residence time of water in a river is analogous to average turnover rate of residents in a large apartment complex. Some may stay only a few months while others live out their lives there, but their average stay tends not to change very much. This number therefore provides an indication of the overall restlessness of the residents as a group.

Rivers typically have average residence times measured in weeks to months, and the average residence for lakes may stretch into centuries. Large lakes located in areas of dry climate tend to retain water the longest; for example, Lake Titicaca on the border between Peru and Bolivia has an average residence of more than thirteen hundred years. Smaller freshwater lakes and ponds typically recycle their water more quickly, on the order of years to decades. Artificial reservoirs built within or near mountainous regions also have short residence times, typically on the order of a few years. Aside from power generation, their primary purpose is to average out seasonal to yearly variations in precipitation, in order to provide a stable water supply.

The ocean is a much larger reservoir than all the world's rivers and lakes combined (Figure 12.1) and it also has a much longer residence time, estimated at about thirty-two hundred years. It fills a global depression caused by the contrast between continental crust, which is relatively thick and light, and oceanic crust, which is relatively thin and dense. To move water out of the ocean basin and onto the continents requires some heavy lifting, which in turn requires energy. As discussed in Chapter 4 this energy originates in the Sun, which evaporates water from the ocean surface and drives it upward into the atmosphere. The trip through the atmosphere is a short one, though, because the amount of solar energy striking the ocean is huge and the amount of water that can be held in air is relatively small. Water resides in the atmospheric reservoir on average only about nine days, the length of a nice vacation. It must then return to the surface as rain or snow.

Around half of the water that falls on the continents returns quickly to the atmosphere, because of evaporation or transpiration (the release of water by the leaves of plants). Most of the rest is ultimately carried to the sea by rivers. However, a small fraction of continental precipitation stays on for a much longer visit, either as glacial ice or as groundwater. The latter consists of water that has percolated deep into sediments or other porous rocks called aquifers. The pores in these sediments or rocks can be thought of collectively as a vast natural plumbing system, which contains the majority of Earth's fresh, liquid water. The depth at which these pores become fully saturated with water is called the water table, a surface that generally mimics the shape of the overlying land surface from which the water has percolated (Figure 12.4). Differences in the elevation of this surface create

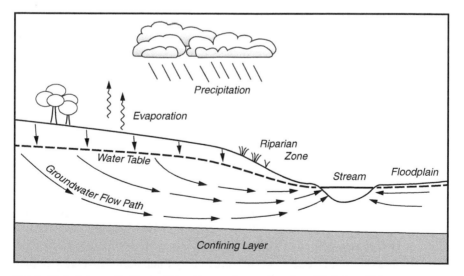

Figure 12.4. Schematic illustration of groundwater flow (modified from U.S. Geological Survey).

pressure differences that generally drive underground flow from areas where the water table is high toward areas where it is low. Rivers and lakes can be thought of as places where the water table is exposed at the surface.

At this point the astute reader may be wondering why we need to worry about the sustainability of water supplies, if the underground reservoir of freshwater is so gigantic. The full answer to this question is complex, but to a first approximation the problem is that groundwater does not flow as rapidly as surface water. It is a slow-shuffling giant, with flow rates measured in meters per year. The practical availability of groundwater therefore depends not only on the overall size of the reservoir, but also on how quickly and efficiently water can be added to or removed from it.

The Speed Limit for Underground Flow

Pipes have been used for water supply since at least the time of the Romans, who constructed them from lead (the English word "plumbing" is from the Latin word for lead, *plumbum*). The pores in sandstone or other rocks can be thought of as a system of miniature pipes, but any plumber who built such a system would probably be fired after his first day on the job. Rather than being straight and smooth conduits that optimize fluid flow, subsurface pores link together to form a tortuous maze of minuscule, labyrinthine passageways. The internal surfaces of these pores are inherently rough and uneven under the best of circumstances and in the worst case may be partly clogged with mineral deposits. Because they are so small (on the order of millimeters to nanometers) pores possess a large amount of surface area compared to their volume. This further impedes flow because of the tendency for a thin layer of water to stick to these surfaces.

Pushing water through this subsurface maze is a little like going through the security line at the airport: It is an inevitably slow business and you just have to resign yourself to the delay. Just how slow depends on the same basic factors that govern flow through pipes, with the added complication of shoddy natural plumbing. The French engineer Henry Darcy worked out the basic relationship governing flow in either pipes or aquifers in the mid-19th century, in what has since become known as "Darcy's law." In simplest terms, Darcy's law says that the rate of flow through a porous substance increases in proportion to the force pushing on it, the cross-sectional area of the flow, and the intrinsic permeability of the sediments or rocks (a measure of the quality of the natural plumbing). It decreases with longer flow paths or more viscous fluids.

Roman plumbing used gravity as the motive force, and plumbing systems were generally designed to carry water downhill. Gravity acting on the water in a closed plumbing system causes its pressure to increase. The primary purpose of municipal water towers, for example, is to ensure that water can be continuously supplied at a constant pressure, which is directly proportional to the height of the tower above the city water mains. Differences in the elevation of the water table achieve the same effect for aquifers. For water to flow there must also be a low-pressure exit point for the system, which in the case of municipal water supply could be an open faucet. The exit point for underground flow occurs where water enters a surface stream or standing body of water.

Note that pressure differences (called gradients) within either a pipe or an aquifer can allow water to flow uphill locally, something that rivers can never do. The use of a garden hose to clean out the rain gutters on the roof of a house provides a simple example of this behavior. Water could in principle flow downward from the city water tower, through the street-level water mains and residential supply line, and then back *upward* again through the hose to the gutters, all as a result of gravity. Naturally flowing or "artesian" wells work by exactly the same principle, except that the pressure-driven flow is confined by impermeable layers of sediment or rock instead of pipes.

The ultimate in permeability could perhaps be represented by a large municipal water main, which is in effect a single smooth, cylindrical pore. Flow rates through such a pore can be quite rapid, if moderate pressure is applied. Geologic materials (rocks and sediments) are all considerably less permeable, but also extraordinarily variable in their ability to permit flow. They span an incredible twelve orders of magnitude in permeability (each order represents a factor of 10)! At the high end are substances such as clean unconsolidated gravels, in which flow velocities may not be greatly slower than in the rivers that originally deposited the gravel. At the low end are rocks such as granite or salt, which can effectively bring flow to a standstill.

As a practical matter, only the most permeable rocks are suitable for long-term groundwater extraction. This situation is similar to production from conventional oil fields, in which high permeability is needed in order for oil to flow. Typical host rocks for water wells include porous sandstone and limestone, but such rocks represent a minority of the Earth's continental crust. The majority of water-saturated rocks have lower permeability, and therefore a much more limited capacity to produce water. Horizontal drilling and hydraulic fracturing could be used to enhance water production from low-permeability rocks, but the market value of water is generally far less than that of oil or gas. The majority of the Earth's fresh groundwater is therefore effectively inaccessible.

Because of the limitations imposed by Darcy's law groundwater flow is much slower than in rivers, and flow through deeper aquifers is generally slower than in shallower ones. This characteristic, combined with the large size of the groundwater "reservoir," makes for residence times that may stretch into the hundreds or even thousands of years! Residence times for deeper brines commonly are even longer, reaching into the millions of years. Such subterranean sluggishness has both advantages and disadvantages with respect to human needs for freshwater supplies. On the good side, groundwater discharge provides an important part of the overall water supply to rivers, sustaining river flow when precipitation and surface runoff are lacking. Natural infiltration of precipitation into floodplain soils and sediments likewise helps to reduce the severity of floods, by delaying the delivery of that water to the river. Slow, consistent groundwater flow also helps protect against seasonal fluctuation of the water level in wells.

On the other hand, wells that have been overpumped take a long time to recover to their original water levels. This has developed into a significant problem in many areas of the world, where local rates of groundwater removal have exceeded rates of natural recharge. The water level in such wells has no choice but to drop, causing a cone-shaped depression of the water table. The associated pressure decrease can result in influx of saline water from deeper brines or from the sea (in coastal areas). Sluggish subsurface flow also means that any pollutants that find their way into a freshwater aquifer tend to stay there a long time and may be difficult or impossible to remove completely.

Fossil Water

The 18th century lexicographer Sam Johnson defined the word "fossil" as anything that is dug from the ground. More recently this term has also become synonymous with nonrenewable natural resources. Groundwater readily meets both definitions; its generally slow rates of natural recharge imply that once depleted, replenishment may require time frames well beyond the life span of living humans. Groundwater extraction thus exploits the leverage of geologic time in much the same way as coal, crude oil, or natural gas.

In agricultural terms this leverage allows areas with only sparse natural vegetation to be transformed into highly productive land, using precipitation that fell in the distant past. For example, irrigation of the San Joaquin Valley of California transformed a semi-desert into a rich agricultural region famous for a variety of fruit and vegetable crops. The greening of marginal lands has a cost, however. Overpumping of groundwater during the mid-20th century quickly depressed water levels in its deep aquifer system as much as 100 m and caused up to 8.5 m of subsidence of the overlying land surface. Improved management, combined with relatively rapid recharge from snowfall in the nearby Sierra Nevada Mountains, have allowed groundwater levels to recover substantially, but the subsidence is likely irreversible.

Irrigation has also allowed agriculture to flourish in the dry central plains of the United States, which possess good soils but marginal rainfall. It has been proposed that native plants such as switchgrass could be grown commercially there and used as an alternative to corn for producing ethanol, potentially lowering the cost and increasing the energy return ratio of biofuels. However, there is a catch. While it is true that grasses grow naturally on the plains, the rate of this biomass production is limited by the availability of water. The plains become progressively drier going west, and net primary productivity at their western edge is only about one-fourth that of the richer parts of the Corn Belt (Figure 12.5). Dry conditions therefore limit the amount of biomass that can be produced per year, and that in turn limits the amount of potential biofuel production.

Luckily the central plains hold another, hidden resource, known as the High Plains Aquifer (Figure 12.6). This massive body of underground water is contained within a complex succession of sedimentary layers that were deposited by east-flowing rivers. Most of these deposits belong to the Ogallala Group, laid down between about 10 and 2 million years ago. The whole package slopes gently eastward, from a surface elevation of about 1,600 m in Denver to 300 m in eastern Kansas. The water-saturated thickness of the High Plains Aquifer exceeds 600 m in places, and its total volume of water storage is roughly equivalent to Lake Huron.

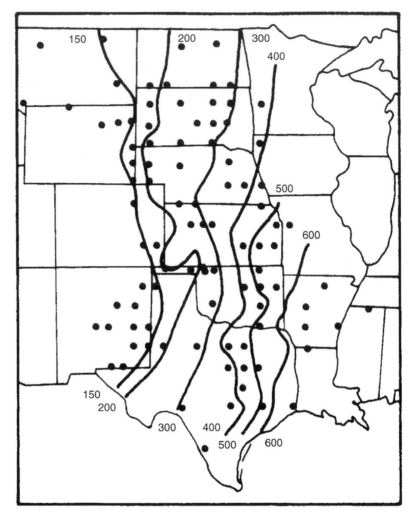

Figure 12.5. Net primary productivity of the central U.S. grasslands during years with average precipitation, measured in grams per square meter per year (Sala et al., 1988).

The High Plains Aquifer in effect serves as a sixth "Great Lake," supplying nearly one-third of the groundwater used for irrigation in the United States. The commercial impact of this water has been dramatic; for example, according to a 2005 study by Christopher Kucharik and Navin Ramankutty of the University of Wisconsin, irrigation in Nebraska, Kansas, and Texas has increased corn yields by

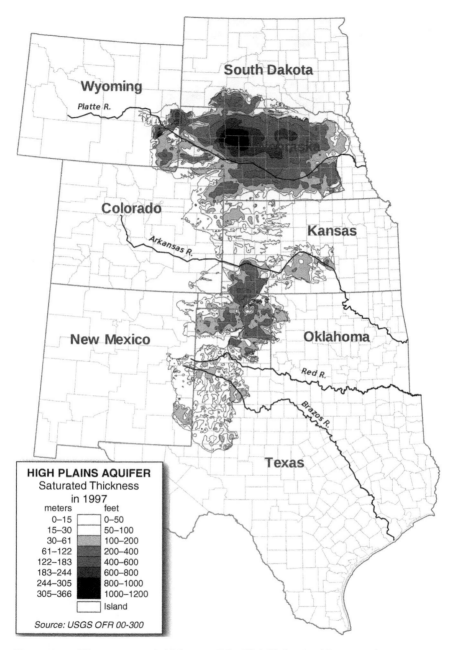

Figure 12.6. Water-saturated thickness of the High Plains Aquifer, central United States (Stanton et al. 2011, U.S. Geological Survey).

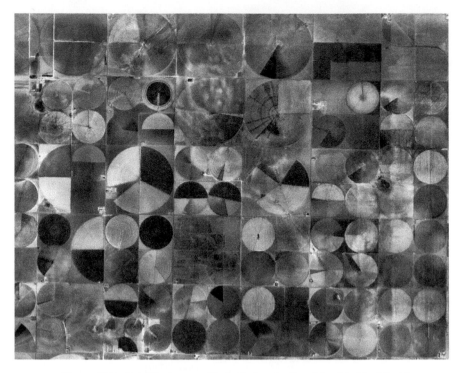

Figure 12.7. Satellite image of irrigated agricultural land in Deaf Smith County, Texas Panhandle.

Each circle represents a field irrigated by ground water applied by a line of sprinklers that pivot around a central point. The small circles in this image are approximately 1/4 mile (400 m) in diameter; the large circles are approximately 1 mile (1600 m) (NASA data).

75–90 percent since the 1950s. The same study noted that irrigation also reduces year to year variability in crop yield by a factor of 3, an important consideration in an area marked by recurring periods of drought. Irrigation is mostly done using center-pivot systems, which create unique patterns of emerald green circles and semicircles super-imposed on a brown backdrop when viewed from above (Figure 12.7). Such patterns typify much of the central plains, and are largely respon-sible for their agricultural productivity.

While Lake Huron and the High Plains Aquifer are similar in volume, Lake Huron is a youthful body of water, with a mean resi-dence time of only about twenty years. The High Plains Aquifer, on the other hand, is a creaky prehistoric relict; radiocarbon ages of water beneath the high plains extend backward as far as fifteen thousand years. The massive continental glaciers of the last ice age were just beginning to retreat at that time, and saber-toothed cats stalked woolly mammoths across the frigid tundra. Most of the water in the High Plains Aquifer arrived later, but its age is still counted in the thousands of years.

Infiltration of new water into the High Plains Aquifer proceeds at a snail's pace, because of a combination of low rates of precipitation; high rates of evaporation on the hot, windy plains; and the presence of relatively impermeable layers of calcium carbonate (called caliche) that have formed near the land surface. These impermeable layers cause much of the precipitation that doesn't evaporate to run off into rivers. According to estimates of the U.S. Geological Survey, the average aquifer recharge rate is less than five centimeters per year, whereas fossil water is commonly applied to irrigated lands at a rate of twenty to forty centimeters per year.

As a result the water table is steadily dropping, and an estimated 8 percent of the original aquifer volume has already been lost. Locally the situation is worse; more than 50 percent of the original saturated aquifer thickness has been lost in parts of Texas and Kansas (McGuire, 2011). Current extraction rates in those areas are likely not sustainable beyond the next few decades (Scanlon, 2012). The situation is better in the cooler, wetter northern plains, particularly in the less densely farmed Nebraska Sandhills region, but this is hardly cause for complacency.

Even regions with apparently abundant freshwater supplies cannot afford to take them for granted. For example, the five North American Great Lakes represent the world's single largest resource of fresh surface water, accounting for about one-fifth of the global total. Despite this watery cornucopia, concerns over worsening water pollution, overconsumption, and drought resulted in an agreement between the United States and Canada that largely prohibits diversion of water outside the Great Lakes watershed. Communities near the Great Lakes but outside their watershed generally must rely on groundwater instead. For example, the western suburbs of Milwaukee get their water from wells that are more than six hundred meters deep. Overpumping of those wells has already caused water levels to drop at rates of up to two meters per year and has created a large cone of depression in the water table. This heavy usage of fossil water cannot be sustained indefinitely.

Water Where You Want It

Real estate agents often assert that the three most important factors determining property values are location, location, and location. To a certain extent the same can be said for freshwater supplies, since they often are not located where we would like them to be. The days of getting our drinking water directly from a local stream or backyard well are long past, at least within the developed world. Some municipalities are able to draw water directly from surface supplies, but virtually all such sources require treatment to render them safe to drink. Treating water requires energy.

As noted in Chapter 4, hydroelectric reservoirs contain gravitational potential energy in proportion to the mass of the water they hold, the height of their dams, and the strength of the Earth's gravitational field. The situation is precisely reversed for water that must be pumped from wells. Gravitational potential energy must be *added* to the pumped water in order to lift it to the surface. The deeper the well, the more energy is required. An alternative approach in regions that lack adequate groundwater supplies is to import water from better-endowed areas, usually mountain ranges, where precipitation rates are much higher. For example, much of the water supply for southern California is transported from the Sierra Nevada Mountains, via a 500-km-long system of aqueducts and pipes known as the California State Water Project.

One might reasonably expect that the water would simply flow through the aqueducts of its own accord, since the Sierra Nevada reach up to 4 km in elevation and Los Angeles lies at sea level. However, the intervening path is obstructed by other, smaller mountain ranges that can rise to elevations of more than 1,000 m. Water must therefore be pumped uphill over these ranges, in order to complete the downhill journey to Los Angeles. Part of the energy expended in pumping water uphill can be recaptured as it flows back downhill, but the inherent inefficiencies of electrical generation, transmission, pumps, and generators mean that much energy is irretrievably lost. As a result, the California State Water Project is the state's largest single consumer of electricity.

Overall it has been estimated that about 3 percent of U.S. electricity consumption is used to obtain water. End uses of water such as heating further add to this total; for example, in 2004, the National Resources Defense Council estimated that water heating and other end uses account for a total of 14 percent of California's electricity consumption and 31 percent of its natural gas consumption.

Water of Last Resort

The most obvious source of water for coastal cities such as Los Angeles is the adjacent ocean, which in principle represents an inexhaustible supply. Its use requires desalination, however, which in turn requires energy. Two different methods of desalination are presently in common use, vacuum distillation and reverse osmosis. The first of these essentially mimics the natural hydrologic cycle, by using heat to vaporize freshwater and thereby separate it from seawater. The second takes a different approach, pumping seawater through a membrane that retains the chemical ions responsible for saltiness. Desalination has traditionally been viewed as a measure of last resort, requiring a capital-intensive infrastructure and large amounts of energy. Desalination is most practical where other alternatives are lacking, for example, on ships at sea or on small islands that lack sufficient groundwater resources. It presently makes its greatest contributions in the Middle East, a region where energy and money are far more abundant than rainfall.

The costs of desalination have steadily declined over the past few decades, however, and the costs of other water supply alternatives have generally risen. Desalination has grown increasingly feasible by comparison. Desalination may be particularly attractive when the alternative is to transport water long distances overland. For example, a 2011 report by the WateReuse Association estimated that the energy required to desalinate water from the Pacific Ocean would be roughly comparable to the energy presently used to transport and treat water obtained through the California State Water Project. Even dry inland areas could potentially use desalination to expand their water supplies. Brackish to saline groundwater (brine) lies beneath much of the interior of the continents, particularly within sedimentary basins. The potential magnitude of this resource is poorly known but certainly enormous. Tapping it would require large monetary and energy investments, however, for drilling, pumping, and desalination.

More fundamentally, desalination represents a significant step backward in the quest to develop renewable energy resources. Enormous quantities of seawater are already desalinated each year by the Sun; humans are required only to steward the resulting existing resource. By expending additional energy supplies to desalinate seawater we turn our backs on renewable solar energy already in hand. While it may be expedient to do so in some cases, it is an intrinsically wasteful solution.

Conclusions

The Earth is remarkably well endowed with water compared to other planets in the solar system, and even though most of this water is either salty or frozen, the remaining amount of freshwater is nonetheless large. Most of this freshwater bounty resides underground, within porous rocks or soil. Because of the inherently slow rates of groundwater flow, high-quality aquifers may be depleted more quickly than they can be replenished, however. The deep aquifers that hold some of the largest resources often require hundreds to thousands of years to recharge and should therefore be considered fossil resources in the same sense as crude oil or natural gas. Slow rates of subsurface flow also mean that any contaminants that find their way into groundwater supplies may remain there for similarly long periods.

Because of its unique physical properties water is used extensively in nearly all forms of large-scale energy production. Biofuel cultivation and distilling consume by far the largest quantities of water per unit of useful energy produced, but other activities such as heavy oil production, oil refining, and hydraulic fracturing also make noteworthy demands. Such uses are encouraged by the fact that in monetary terms, water is far less costly than the fuels it helps to produce. For example, at oil prices and residential water rates prevailing in the United States at the time of this writing, approximately five hundred to one thousand barrels of water can be bought for the cost of one barrel of oil. Irrigation water rates are even cheaper; for example, a 2010 report written by Dennis Wichelns for the Organization for Economic Cooperation and Development estimated that farmers in the Central Valley of California paid at most about $103 per 1,000 cubic meters. This works out to a cost ratio of around five thousand barrels of water to one barrel of oil.

The low monetary price paid for water does not necessarily reflect its true intrinsic value, however. Wisconsin hydrogeologist Ken Bradbury summarized the situation nicely with the question "If marooned on a desert island which would you rather have, a barrel of oil or a barrel of water?" This scenario is complicated, however, by the fact that we also need energy to obtain and use water. That hypothetical barrel of water did not simply appear on the island; it had to be pumped out of the ground, desalinated, or carried there by boat. Mainland water use in semiarid to arid regions may also require long-distance pumping through pipes or aqueducts. More energy still is needed if we would like to heat water to cook, to wash clothes, or to take a hot shower. Water and energy are therefore inextricably linked.

For More Information

Alley, W. M., 2003, Desalination of Groundwater: Earth Science Perspectives: *U.S. Geological Survey Fact Sheet* 075-03, 4 p.

Anonymous, 2006, *Energy Demands on Water Resources: Report to Congress on the Interdependency of Energy and Water*: Washington, D.C., U.S. Department of Energy, 80 p.

Anonymous, 2008, Water Implications of Biofuels Production in the United States: Washington, D.C., National Research Council of the National Academies, 76 p.

Anonymous, 2011, Seawater Desalination Power Consumption: Alexandria, Virginia, United States, WateReuse Association, 16 p.

Cech, T. V., 2005, *Principles of Water Resources – History, Development, Management, and Policy* (2nd ed.): New York, John Wiley and Sons, 468 p.

Chiu, Y.-W., Walseth, B., and Shu, S., 2009, Water embodied in bioethanol in the United States: *Environmental Science and Technology*, v. 43, p. 2688–2692.

Cohen, R., Nelson, B., and Wolff, G., 2004, *Energy down the Drain: The Hidden Costs of California's Water Supply*: Oakland, California, National Resources Defense Council and the Pacific Institute, 78 p.

Dutton, A. R., 1995, Groundwater isotopic evidence for paleorecharge in the U.S. high plains aquifers: *Quaternary Research*, v. 43, p. 221–231.

Gleick, P. H., 1994, Water and energy: *Annual Review of Energy and the Environment*, v. 19, p. 267–299.

Kenny, J. F., Barber, N. L., Hutson, S. S., Linsey, K. S., Lovelace, J. K., and Maupin, M. A., 2009, Estimated use of water in the United States in 2005: *U.S. Geological Survey Circular* 1344, 52 p.

Kucharik, C. J., and Ramankutty, N., 2005, Trends and variability in U.S. corn yields over the twentieth century: *Earth Interactions*, v. 9, p. 1–28.

McGuire, V. L., 2011, Water-level changes in the High Plains Aquifer, predevelopment to 2009, 2007–08, and 2008–09, and change in water storage, predevelopment to 2009: *U.S. Geological Survey Scientific Investigations Report* 2011-5089, 13 p.

National Energy Board of Canada, 2006, *Canada's Oil Sands, Opportunities and Challenges to 2015: An Update*: Calgary: National Energy Board of Canada, , 71 p.

Quinn, F. H., 1992, Hydraulic residence times for the Laurentian Great Lakes: *Journal of Great Lakes Research*, v. 18, p. 22–28.

Sala, O. E., Parton, W. J., Joyce, L. A., and Laurenroth, W. K., 1988, Primary production of the central grassland region of the United States: *Ecology*, v. 69, p. 40–45.

Scanlon, B. R., Faunt, C. C., Longuevergne, L. Reedy, R. C., Alley, W. M., McGuire, V. L., and McMahon, P. B., 2012, Groundwater depletion and sustainability of irrigation in the U.S. High Plains and Central Valley: *Proceedings of the National Academy of Science*, v. 109, p. 9320–9325.

Stanton, J. S., Qi, S. L., Ryter, D. W., Falk, S. E., Houston, N. A., Peterson, S. M., Westenbroek, S. M., and Christenson, S. C., 2011, Selected approaches to estimate water-budget components of the high plains, 1940–1949 and 2000–2009: *U.S. Geological Survey Scientific Investigations Report* 2011–5183, 79 p.

Valley, J. W., 2005, A cool early Earth?: *Scientific American*, October, p. 58–65.

Wichelns, D., 2010, *Agricultural Water Pricing: United States*: Paris, Organization for Economic Co-operation and Development, 27 p.

Zektser, I. S., and Everett, L. G., eds., 2004, Groundwater Resources of the World and Their Use: Paris, UNESCO, 346 p.

13

Primordial Power: Geothermal and Nuclear

At daybreak on July 4, 1054, a Chinese astronomer named Yang Weide noted the appearance of a new "guest star" in the eastern sky, near the more familiar reference star Tianguan (now known as Zeta Tauri). The appearance of a guest star was an exceedingly rare event considered to carry great geopolitical significance; the emperor Renzong therefore was duly notified. The new star remained visible during the daytime for 23 days, and finally disappeared from the night sky after 642 days. At the time, the Song dynasty was a world technological leader, having introduced paper money, gunpowder, and use of the magnetic compass for marine navigation. They did not yet have the telescope, however, and so could not know that the new guest star had been replaced by a much dimmer but still luminescent interstellar cloud, the Crab Nebula (Figure 13.1).

Yang's guest star was in fact a supernova, which has since been named SN 1054. It formed by the collapse of an aging star, which originally had a mass many times that of our own Sun. As it exhausted its core fuel supply, the star eventually reached a point where the energy released by nuclear fusion could no longer counteract the unimaginably large gravitational forces pulling inward. A violent implosion ensued, immediately followed by an explosion that temporarily made SN 1054 one of the brightest objects in the galaxy. Within seconds, the energy of this explosion also spawned a suite of new chemical elements, which were immediately hurled outward into interstellar space. Among them was one of the rarest naturally occurring elements of all, uranium.

Born of extreme violence, uranium remains inherently unstable or "radioactive" for the rest of its life, fated to spontaneous and inexorable decay to lighter elements. Some of these second-generation or "daughter" elements are themselves unstable, experiencing further

Figure 13.1. The Crab Nebula (NASA Hubble Space Telescope).

decay until the process eventually reaches its end with drearily dull but reliably stable lead. Along the way, the explosive energy that was originally captured in seconds is slowly released over periods that stretch into the billions of years. The rate of this natural decay is often expressed in terms of "half-life"; one half-life is the time a particular isotope takes to lose half of its original atoms (isotopes are variants of an element that differ in their number of neutrons). The half-life of the most common isotope of uranium (^{238}U) is about 4.5 billion years, approximately equal to the age of the Earth itself.

The nuclear potential energy stored in uranium and other radioactive elements can be put to practical use, but some daunting obstacles must be overcome to do so. To start with, we have no practical way to retrieve uranium that has been expelled from supernovae located hundreds of light-years away from Earth. Even if we could somehow travel such distances, the extremely low density of supernova remnants would make direct harvesting difficult if not impossible.

Fortunately, the extremely long half-lives of ^{238}U and other radio-active isotopes afford a natural solution to this problem. Rather than visit far distant nebulae, we can instead make use of the remnants of supernovae that exploded in our own galactic neighborhood, a very long time ago. Some of this material was incorporated into the Earth during its original formation. We can exploit it as an energy source in two distinctly different ways. First, we can tap into the geother-mal heat generated by natural decay of radioactive isotopes within the Earth, primarily ^{238}U, ^{235}U, ^{232}Th (thorium-232), and ^{40}K (potassium-40). Second, we can mine uranium and then artificially break it down into smaller atoms via bombardment with neutrons, a process known as fission. This is the basis of all nuclear power production today.

Warmed from Within: The Interior Heat of the Earth

Jules Verne's 1864 novel *Journey to the Center of the Earth* depicts a fan-ciful subterranean world containing giant mushrooms, prehistoric monsters, and even an underground sea. Although a good read, this novel ignores the fact that temperatures near the center of the Earth would quickly incinerate any living organism. Recently published cal-culations put the temperature at the boundary between solid inner core and liquid outer core at approximately 6000°C, which would vaporize iron were it not for the enormous pressures found there. In contrast, the Earth's surface averages only about 15°C. The huge tem-perature difference between its center and its surface allows the Earth to act as a giant heat engine, converting thermal energy into kinetic energy with a maximum theoretical efficiency in the range of 95 per-cent (see the discussion of Carnot efficiency in Chapter 3). This heat engine moves the tectonic plates, by inducing large vertical move-ments of mantle rocks. It also helps to generate the Earth's magnetic field, which results from a combination of convective currents and rotational movement in the electrically conductive outer core.

About half of the Earth's internal heat is left over from its ear-liest days, when the whole planet was molten or nearly so. This melt-ing occurred in part through the collisions of protoplanets and other space debris moving at high velocity; their kinetic energy was con-verted into heat energy. The same phenomenon can be easily demon-strated by repeatedly striking a rock with a hammer, a process that if continued long enough makes both the hammer and the rock feel warmer. Radioactive decay of short-lived isotopes such as ^{26}Al and ^{60}Fe also helped to heat things up during the first few million years.

Once the Earth was largely molten, liquid iron sinking toward its core administered the coup de grâce, by converting its gravitational potential energy into heat.

As quickly as the Earth heated up it began to radiate its heat into the cold emptiness of space, not unlike a northern lake that loses heat to the frigid air of an approaching winter. The lake will soon form a thin crust of ice, and the Earth likewise formed a thin crust of rock. The mantle solidified soon afterward (in geologic terms) but remains blast furnace hot, with present-day temperatures typically in the range of 1000°C –3000°C. This raises an interesting question: How has the Earth's interior managed to remain so warm, more than 4 billion years after its formation?

Part of the answer is that rocks are pretty good insulators. This insulating quality, which can be expressed as thermal conductivity, measures how quickly heat energy can be transferred from one place to another. Metals have very high thermal conductivity (Figure 13.2); that is why you do not want to grab the edge of a metal cooking pan that is being heated on a stove. The thermal energy of the stove quickly spreads from the bottom of the pan to its edges. At the other extreme, air has very low thermal conductivity and therefore is an excellent insulator, provided it remains stationary. Goose down insulation in clothing takes advantage of this property; it is not the goose down itself that keeps you warm, but the large volume of air trapped motionless between its fibers.

Rocks do not insulate as well as air, but they are about 100 times better insulators than metals. The Earth's rocky crust therefore functions as a sort of planetary overcoat, which greatly slows the loss of heat to space. The situation in the mantle is a bit different. As noted in Chapter 2, mantle rocks migrate slowly upward and downward, carrying heat from the molten, metallic outer core with them as they do so. This convection raises temperatures in the uppermost parts of the mantle well above what would be expected for a static, cooling Earth. The Earth's crustal overcoat therefore experiences a surprisingly large temperature gradient over a relatively short distance.

Glowing in the Dark: Nuclear Decay inside the Earth

The other half of the Earth's internal heat results from nuclear decay of relatively long-lived radioactive isotopes that were incorporated

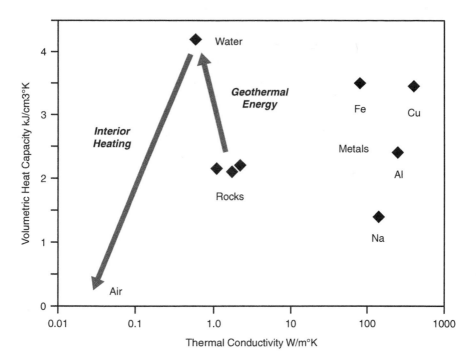

Figure 13.2. Comparative thermal properties of different materials (note that each division on the horizontal scale indicates a tenfold difference in thermal conductivity).

Geothermal energy extract takes advantage of the high heat capacity of water to transfer the heat contained in rocks to the surface (arrow pointing up). There it can be flashed to steam to drive an electrical generator (if sufficiently hot), to supply domestic hot water, or to warm the air inside buildings (arrow pointing down).

during its original formation. Nuclear decay is the spontaneous breakdown of atomic nuclei, which releases some of the energy that originally bound those nuclei together. The products of this decay have three basic forms: alpha particles (α), beta particles (β), and gamma rays (γ). Alpha particles are nothing more than helium atoms without their electrons; they consist of two protons and two neutrons. In fact, nuclear decay is the principal source of commercially available helium. The lightweight stuff that fills party balloons and makes your voice sound like Donald Duck therefore started out as dense uranium (or other radioactive isotopes) and was obtained by drilling wells deep into the Earth's crust.

Beta particles are otherwise known as electrons. It might seem a bit strange at first that electrons can be emitted from an atomic nucleus, which is nominally composed only of protons and neutrons. The creation of beta particles is made possible through a bit of sub-atomic sleight of hand, wherein a neutron is converted into a proton and an electron is emitted in the bargain. The resultant nucleus maintains essentially the same mass as it had before but has gained a proton and therefore becomes a different chemical element. Beta decay also releases an antineutrino, which is an electrically neutral particle that can travel through rocks virtually unimpeded. Measurement of the flux of antineutrinos near the Earth's surface provides us with a direct measure of the magnitude of radioactive heating.

Alpha and beta particles both possess the ability to cause cellular damage in human tissues, but for alpha particles this risk is fairly minor because of their inability to penetrate beyond the uppermost layers of skin. Beta particles go deeper and can potentially pose a threat to health, but they can be easily blocked by a thin layer of metal. If desired, you can avoid all risk from beta particles simply by wrapping yourself head to toe in aluminum foil. Gamma rays are a form of electromagnetic energy, similar to x-rays except higher in frequency. Like x-rays they can easily penetrate human tissues, potentially leading to radiation sickness, cancer, or even death. They can only be blocked by massive shielding, such as relatively thick layers of lead.

Fortunately the rocks of the Earth's mantle and crust absorb most of the particles and gamma rays associated with nuclear decay, preventing us from being fried by natural radiation. As it is absorbed by rocks, this energy is converted into geothermal heat, in proportion to the abundance of radioactive isotopes and the rates at which they decay. Four parent isotopes do most of the heating: ^{238}U, ^{235}U, ^{232}Th, and ^{40}K. Each one ultimately decays to a stable daughter element, but getting to that point can be a surprisingly complicated affair when viewed in detail. For example, the decay of ^{238}U progresses along a tortuous path marked by eighteen other unstable isotopes, before finally arriving at stable lead (Figure 13.3). These intermediaries have vanishingly short half-lives compared to ^{238}U's, but they can be noteworthy despite their transience. Radon gas (^{222}Rn), for example, has been shown to cause lung cancer if breathed in sufficient quantities. Its short half-life of only 3.8 days normally mitigates this risk, but problems can arise in closed spaces such as basements, where radon emanating from underlying rocks might become concentrated in the stagnant air supply.

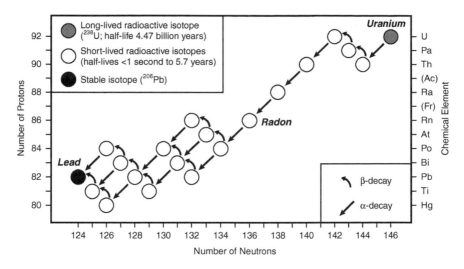

Figure 13.3. Natural radioactive decay sequence of ^{238}U.

Hot Rocks: Heat from the Earth

The insulating qualities of the Earth's crust are wonderfully convenient for allowing us to walk about on the surface without burning our toes. However, for the purpose of geothermal energy extraction these insulating qualities are perhaps a bit too good; the globally averaged upward heat flow is only about 87 milliwatts per square meter. Even if this minuscule heat flow could somehow be recovered at 100 percent efficiency (it can't), it would take an area about the size of a tennis court to power a single 20 watt compact fluorescent light bulb. In contrast, the Earth absorbs solar energy at a rate that is roughly two thousand times greater, roughly 168 watts per square meter on average. The ambient temperature of the Earth's surface rocks and soil therefore depends very much on incoming sunlight, and very little on the hot interior below.

Because the Earth is very large, the total geothermal heat flow rate is still an impressive number, estimated at between 44 and 47 terawatts. This is equal to roughly three times the global power consumption of humans in the early 21st century. However, only a small fraction of this power can ever be put to practical use. For starters, about two-thirds of the total is accounted for by midocean ridges, which lose heat at an average rate around 300 mW/m^2 and act as exhaust pipes for the Earth's internal heat engine. The ridges are cooled by a continuous stream of seawater flowing through them and thus serve as a cold sink against which the hot interior can operate. Midocean ridges by definition lie far from the continents, a distance that makes their practical use by land-dwelling humans rather inconvenient. Exceptions might perhaps be made for places where midocean ridges terminate near a continent, for example, where the East Pacific Rise abuts Mexico. At present, however, the only significant use of ridge heat occurs in Iceland, a unique volcanic island that is perched precariously atop the Mid-Atlantic Ridge.

Heat flow from the continents averages only about 50 mW/m². Furthermore, continental geothermal heat is unevenly distributed, and not always found where it is most needed. The greatest heat flows occur above subduction zones and in areas where hot mantle rocks rise up beneath the continents. In the United States this limits most of the better geothermal prospects to the western third of the country. The Basin and Range Province, a sparsely populated region of mountains and high deserts that is centered on Nevada, holds much of the U.S. potential, along with a few other areas such as Yellowstone, The Geysers, and the Imperial Valley. Across the Atlantic the prospects are generally best in southern Europe, especially near the Mediterranean coast. The entire Pacific Rim and eastern Africa also experience enhanced heat flows. In contrast, the crust in tectonically quiescent areas such as the central and eastern United States and northern Europe is relatively cool.

Even in the places where heat flow rates are highest, they are still far too small to satisfy any practical need. Geothermal energy therefore is not directly based on the rate at which heat is currently flowing to the surface, but instead on the amount of heat that has been *stored* in hot rocks. A simple comparison might be made to a large lake that is fed by a small stream and drained by another exit stream. The stream flow into the lake, which is analogous to the Earth's heat flow, is very modest. Over long periods, though, it manages to fill the entire lake to its spillover point. The water in the lake is analogous to the heat content of the Earth's rocks. Note that in both cases the resource can also become depleted, if it is withdrawn faster than it is replenished.

Strictly speaking, geothermal energy should therefore be considered a nonrenewable resource in the same sense as coal, crude oil, or natural gas. Its practical use depends fundamentally on extracting heat energy at rates much faster than natural replenishment. Ideally this requires that the rocks be as hot as possible, maximizing both their heat content and the rate at which heat can flow out. The hottest accessible rocks are found in and around volcanoes, which are places where nature has helped to bypass the Earth's insulating crust by rapidly transporting molten rocks (magma) from great depth upward to the surface.

Volcano Power

In 79 AD Mount Vesuvius, located near the modern Italian city of Naples, exploded in a cataclysmic eruption that shot ash upward into the stratosphere and buried nearby Pompeii and Herculaneum along with most of their inhabitants. The Roman author Pliny the Younger witnessed the eruption safely from across the Bay of Naples, but his uncle (also named Pliny) died while attempting to rescue survivors. Pliny the Elder's death should not have been a complete surprise, considering that a gigantic eruptive cloud visible for hundreds of kilometers was actively dropping massive quantities of rocky debris onto the immediate area. However, the most deadly phase of the eruption struck without warning. In a letter written to the senator and historian Tacitus, Pliny the Younger recalled how "on Mount Vesuvius broad sheets of fire and leaping flames blazed at several points, their bright glare emphasized by the darkness of night." He was likely describing pyroclastic flows, which are red-hot volcanic avalanches that rush down the sides of volcanoes at speeds rivaling those of jet aircraft. The temperatures in the pyroclastic flows on Vesuvius have been estimated at 250°C–500°C.

Eruptions such as Mount Vesuvius or Mount St. Helens illustrate geothermal energy in its most violent form, though not at its most useful. Dramatic as they are, such eruptions do not even account for very much of the Earth's geothermal heat flow. British geologist David Pyle estimated in 1995 that they represent only about 1 percent of the total amount of heat reaching the Earth's surface. Plinian style eruptions therefore contribute far more to volcanic hazards than to energy resources. This does not mean that the hot rocks associated with volcanoes cannot be put to practical use; indeed they can. To do so, however, requires that volcanic activity be dormant, less explosive, or that it be both.

A verdict of dormancy is based on the long-term absence of obvious evidence such as earthquakes, ground tilting, or steam rising up from the ground. Absence of evidence is not evidence of absence, though. We know from the geologic record of ash and other deposits that long periods of volcanic boredom are commonly interrupted by brief moments of fiery terror. The boredom may last from hundreds to hundreds of thousands of years, which in human terms might make a volcano seem as good as dead. In geologic terms, though, it may just be "playing possum," ready to spring back to life quickly when no one is looking. The Yellowstone volcanic system in the United States appears to exemplify this sort of behavior. It has erupted several times over the past 2 million years, with the largest of these eruptions estimated to have spewed twenty-five hundred times as much ash as the 1980 eruption of Mount St. Helens! Yellowstone is not finished yet, although current geologic consensus holds that it is unlikely to project another violent eruption during the next ten thousand years or so.

The safest strategy is to exploit volcanoes that are not predisposed to such violence in the first place. These are common enough; for example, Iceland and the Hawaiian Islands are built mostly from slow-moving, benign lava that hardens into sheetlike layers of volcanic rock called basalt. One key to such calm volcanism is to start with a magma that contains relatively little water or volatile gases, bubbles of which can propel violent eruptions. Another key is that the magma be relatively thin and runny in consistency, so that any gas bubbles that are present can more peacefully escape. Lavas erupted from these milder volcanoes may still destroy buildings in the path of their inexorable slow flow, but rarely do they kill.

Drilling Deeper: The Geothermal Gradient

Both volcanoes and oil fields represent high-quality energy resources that are highly localized in nature. Active volcanoes are considerably easier to find, of course, particularly adjacent to the Pacific Ocean, the Mediterranean Sea, and a handful of other overheated locales. Those not lucky enough to have their own volcano can still exploit geothermal energy, however, provided they are willing to drill downward to find it.

If the Earth's crust acts as a planetary overcoat, you would expect to find a large temperature contrast between relatively cool rocks near the surface and the hot rocks deeper in the crust. In the case of an actual overcoat, the human body is regulated at about 37°C and on a frigid January day the outside temperature might be –10°C, a difference of nearly 40°C across 2 cm of fabric. If you poked a thermometer slowly through the coat you might expect to find that temperature increases inward at a rate of about 2 degrees per millimeter.

Although the numbers are different for the Earth's crust, the basic scenario is analogous, with one important difference: The crust itself produces heat due to radioactive decay of isotopes contained in its rocks. Rather than an overcoat, the crust therefore behaves as more of an electric blanket. As a result, the greatest temperature contrasts are found relatively near the surface. Temperatures in the upper 5–10 km of continental crust increase about 25°C for every kilometer of increased depth on average. To find temperatures hot enough to boil water you would therefore have to drill down about 4 km. Below the crust temperatures continue to increase, but at a slower rate.

A big advantage of tapping deep geothermal heat is that it dramatically expands the total amount of energy that might be recovered. For example, according to a 2006 report by the Massachusetts Institute of Technology, sedimentary basins in the United States hold a geothermal resource that is comparable in magnitude to volcanoes. The igneous and metamorphic rocks that make up most of the continental crust were estimated to hold more than one hundred times as much geothermal energy as either volcanoes or sedimentary basins.

Hot Rocks and Water

Hot rocks by themselves frankly are not good for much. To heat your house you would have to pick them up and carry them inside, a cumbersome and sweaty endeavor at best. You could perhaps save some effort by lying down directly on the warm crust of a recently solidified lava flow instead, but you had better mind the thin spots! There is no practical way to make electricity directly from hot rocks, as can be done with sunlight, and hot rocks clearly cannot be made to flow through a turbine.

What we really want is hot water, or better yet, steam, either of which is vastly more useful. Water is easy to move, cheap, and readily available. Its thermal conductivity is about the same as that of rocks. However, because of its unique molecular structure the volumetric heat capacity of water is about twice that of a similar volume of rock. This makes water remarkably efficient for moving heat from one place to another. And water warmed to 100°C at sea level atmospheric pressure boils and turns into steam, with a large volume increase that can be used to drive a turbine. It's no wonder that water is the universal heat transport medium of choice, both in nature and in geothermal energy development. A variety of schemes have been concocted for

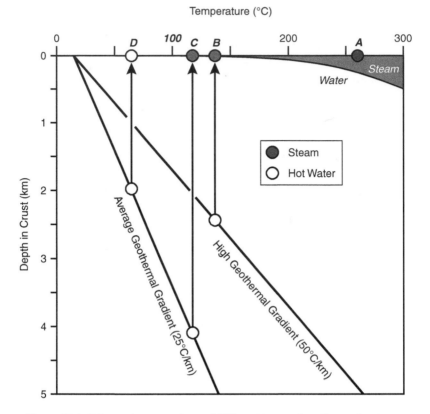

Figure 13.4. Schematic comparison of different types of geothermal systems, based on their depth and temperature.

using water to exploit geothermal heat, some more successfully than others. To a first approximation the effectiveness of these systems can be graded according to the naturally occurring rock temperatures involved, combined with some consideration of how readily water can be made to flow through those rocks.

Leading the class with a letter grade of "A" are systems that exploit natural reservoirs of steam or superheated water located at relatively shallow depth (Figures 13.4 and 13.5). These "hydro-thermal" reservoirs are heated by volcanic rocks, and temperatures can reach several hundred degrees Celsius. Some of the most pro-lific geothermal systems in the world tap into hydrothermal res-ervoirs, including the Ladararello area in northern Italy, which is home to the world's first commercial geothermal power plant, and

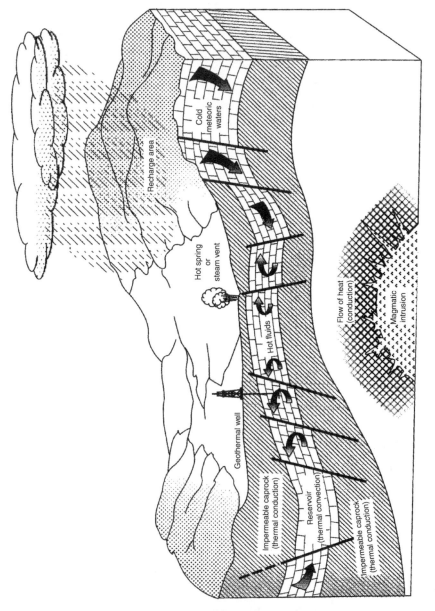

Figure 13.5. Schematic illustration of a hydrothermal system. (U.S. Geological Survey).

the Geysers field in northern California, the world's most prolific geothermal field. The reservoirs for Ladararello and The Geysers consists of porous and permeable sedimentary rock through which water and steam can flow. Similarly porous and permeable pyroclastic flows form the main reservoirs for the Wairakei Power Station in New Zealand.

Systems based on continental geothermal gradients rather than localized volcanic activity generally get lower grades, both because they typically produce lower temperatures and because they require drilling deeper, more expensive wells. A grade of "B" is assigned to systems developed in areas with higher than average geothermal gradients, which mean that they can reach rocks hotter than 100°C by drilling only 2–3 km down. Grade "C" goes to areas with average geothermal gradients, where 100°C temperatures are found at or below 4 km depth. Although these temperatures may be high enough to generate steam and drive a turbine, the cost of generating electricity will be substantially higher than for hydrothermal systems because of the need for deeper drilling.

A grade of "D" is reserved for cases where the rocks encountered by drilling remain below 100°C. Steam turbines are not an option, although electricity may still be generated using a binary system in which hot water is used to heat an organic fluid, such as isobutane, that has a lower boiling point. At present the use of such resources is mostly restricted to domestic hot water supply and heating rather than electricity generation.

Geothermal Plumbing

Regardless of their assigned letter grades, all of the preceding scenarios depend on maintaining continuous flow of water or steam through a natural underground network of pore spaces and fractures. As such, they are ultimately governed by Darcy's law (see Chapter 12), which limits the rate at which groundwater, and the heat it carries, can travel. This limit applies both to the rate at which heat can be extracted and to the rate at which the groundwater that sustains the geothermal system can be replenished.

Relatively shallow high-temperature systems commonly receive natural groundwater recharge from precipitation on the surrounding landscape (Figure 13.5). Natural recharge may not always be fast enough, however; for example, The Geyser's geothermal field

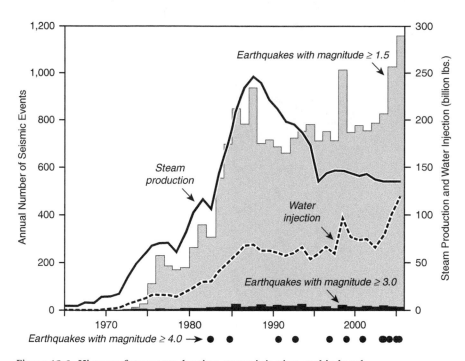

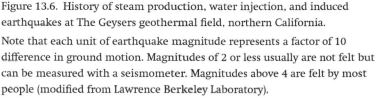

Figure 13.6. History of steam production, water injection, and induced earthquakes at The Geysers geothermal field, northern California.

Note that each unit of earthquake magnitude represents a factor of 10 difference in ground motion. Magnitudes of 2 or less usually are not felt but can be measured with a seismometer. Magnitudes above 4 are felt by most people (modified from Lawrence Berkeley Laboratory).

experienced a rapid decline in production during the late 1980s to early 1990s (Figure 13.6). Injection of recycled municipal wastewater and better management practices slowed the rate of its decline by the late 1990s. Other geothermal projects have fared better; the Wairakei Power Station has maintained production at or above its initial peak for nearly fifty years.

Natural plumbing issues become even more critical for deep wells drilled to exploit the continental geothermal gradient. Surface waters must travel much farther into the subsurface to replenish depleted aquifers and may do so only very slowly if at all. In addition to their heat content, deep brines may contain substantial amounts of dissolved natural gas, a potentially significant energy resource in its own right.

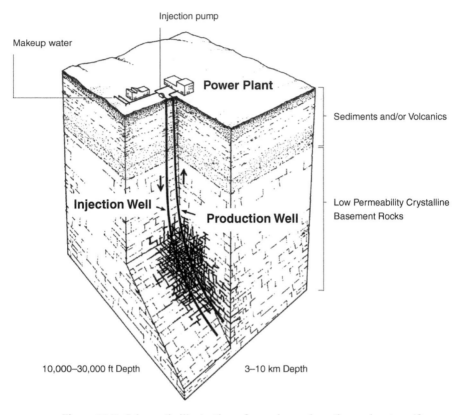

Figure 13.7. Schematic illustration of an enhanced geothermal system (from Tester et al., 2006).

Unfortunately the igneous and metamorphic rocks that hold the lion's share of subsurface heat also contain very little water, because of their dense and impermeable nature. Such rocks share much in common with the gas- and oil-bearing mudstone discussed in Chapter 10: They hold a plentitude of energy but strongly resist its extraction. Logically enough, hydraulic fracturing techniques similar to those used to extract gas from shale can also provide a means of getting heat out of granite. Such "enhanced" geothermal systems operate by injecting cold water down a well at elevated pressure, which then opens and propagates existing natural fractures and induces new fractures to form (Figure 13.7). The injected water is heated as it travels through these fractures, and then extracted through a second well.

Geothermal in Your Back Yard

"Geothermal" also encompasses an entirely different sort of system, the basic principle of which has been known at least since Paleolithic humans began painting the walls of caves. Those early artists would have noticed that the ambient temperature within deep caves changes little during the year, regardless of how the weather above may change. For example, the air in Carlsbad Caverns in New Mexico remains a cool 13°C, even though typical outside temperatures may vary from below 0°C to greater than 40°C. You can investigate this phenomenon by simply walking downstairs to your own basement. Rocks and soil are good insulators, slow to heat during the summer and slow to cool in the winter.

Home "geothermal" systems take these simple principles a step further, by circulating water through a network of pipes buried outside. The simplest application of this system would be to pump water continuously through a heat exchanger in your house, and then back outside again. The heat exchanger is like a reversible radiator, which passively transfers heat from the water to the colder air inside the house during the winter, and from the relatively warm air back into the water during the summer. Note that in both cases the final air temperature still needs some adjustment, however. The very warmest temperature that could be directly obtained from this system would be that found a few meters below the ground surface. Temperatures there remain a cool 13°C–15°C year-round, a bit too chilly for comfortable indoor living.

Home geothermal systems must therefore be supplemented through the inclusion of a heat pump, a sort of thermal amplifier that uses electricity to force additional heat to flow "uphill" from a cooler area (the ground) to a warmer one (your home). A refrigerator provides a common example of a heat pump; it forces heat to flow from its cold interior toward the warmer air of the kitchen. If you could turn your refrigerator inside-out, it would instead heat its interior and cool the kitchen. A larger heat pump can do the same for your entire house. The cost of the extra heat needed to "top up" household temperatures in the winter shows up on your electric bill; you could in fact use electric space heaters to achieve exactly the same effect. However, heat pumps have the added advantage of also being able to reverse their operation in the summer to help with cooling. Heat pumps do not really tap into the interior heat of the Earth, as described earlier in this chapter, but they do help to reduce the amount of energy that must be obtained from other sources.

Geothermal Made Faster: Nuclear Fission

The Earth's internal reservoir of heat is clearly enormous, but the inherently slow rates at which heat and water are able to flow through rocks impede its use. What's needed is a way to speed matters up. Unfortunately there is little we can do about the heat and fluid flow properties of rocks, and therefore no way to accelerate the natural release of heat from the Earth's deep interior. However, there *is* something we can do to speed up the release of nuclear energy from radioactive elements found in the Earth's crust.

Natural radioactive decay chips away at large unstable atoms a little bit at a time, producing a slow but predictable stream of small subatomic particles and correspondingly modest amounts of energy. Faster release of nuclear energy requires more violent intervention, by bombardment of large atomic nuclei with neutrons. If a neutron can be made to strike the nucleus of an already unstable isotope such as ^{235}U at the right velocity, the nucleus will swell by one mass number, then quickly split apart into two smaller atoms (Figure 13.8). This fragmentation is known as fission, which derives from a Latin word that means "to split." Atoms that can be readily split are called fissile, and the smaller atoms that result are called fission products.

If you could weigh a ^{235}U atom before fission and weigh its fission products and leftover neutrons after, you would find that a small

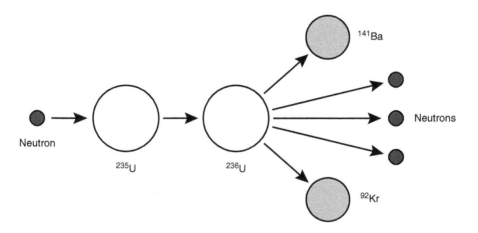

Figure 13.8. Schematic illustration of ^{235}U nuclear fission.
^{141}Ba and ^{92}Kr are typical fission products, with atomic masses substantially less than uranium. Note that three neutrons are released in the process, which may cause fission of other uranium nuclei.

amount of mass, equal to about 0.1 percent of the total, had been lost in the process. This small amount of mass is converted into a large amount of energy, which mostly takes the form of kinetic energy imparted to the fission fragments. The energy released by fission of a single uranium atom is rather small, but the two to three additional neutrons released in the process can strike other ^{235}U atoms, causing additional fission reactions. More help is obtained from the fission products, which are themselves unstable and yield more neutrons as they quickly decay.

The likelihood that any of these newly liberated neutrons will strike other uranium nuclei depends both on how many such targets are available, and how densely they are packed together. If uranium atoms are too few in number or arranged too far apart, fission will grind to a halt. If, however, uranium atoms are abundant and closely packed, the neutrons may trigger a chain reaction in which the rate of fission events rapidly increases through time. Nuclear weapons take this principle to its explosive extreme. The ideal condition for nuclear power generation occurs when an average of one neutron released by each fission event triggers another fission event, resulting in a steady state in which the rate of fission remains constant. The packing configuration of uranium atoms needed to obtain this stable situation is known as critical mass. In nuclear reactors the rate of reaction can be adjusted by moderating the flux of neutrons, using either water or specially formulated control rods.

The total of energy that can be quickly released by nuclear fission is staggering; for example, complete fission of 1 kg of pure ^{235}U will yield the energy equivalent of roughly 2 million kg of crude oil. Because uranium's density is about twenty times that of oil, its volumetric energy density is about 40 million times greater! A kilogram of uranium is only a bit larger than a golf ball but holds the same energy potential as an Olympic-size swimming pool filled with crude oil.

Like high temperature geothermal systems, nuclear fission reactors typically transfer their heat to water, which can then be used to drive a steam turbine. They also enjoy some unique advantages over natural systems. Rather than flowing slowly through a tortuous maze of underground pores and fractures, the water in a nuclear reactor can be pumped quickly through smooth metal pipes. Reactors can also be maintained at uniformly high temperatures, which can be adjusted at will. The net result is a relatively efficient system for quickly converting primordial energy into useful electricity.

Rising to the Top: Uranium Enrichment in the Earth's Crust

There's one small obstacle that must be overcome in order to exploit nuclear fission as an energy source: We must somehow collect enough uranium, one of the rarest chemical elements in the universe, to make a nearly pure reactor fuel. Since the average concentration of uranium in the Earth is about 20 parts per *billion,* making pure uranium fuel might seem about as easy as building a skyscraper out of diamonds. We need the uranium found in rocks to be concentrated by a factor of about 50 million, a task that would be utterly impossible to complete at any meaningful rate. Fortunately the Earth itself has already done most of the work for us, however, plugging away steadily for the past few billion years.

The job began with the early formation of the presolar nebula, a wispy cloud composed mostly of hydrogen and helium but that also included uranium and other heavy elements. Most of the cloud was used up in making the Sun itself, while the remainder formed a large rotating disc. Over time this disc grew progressively "lumpier" by the accretion of small particles into larger ones, culminating with formation of the planets that we know today. Uranium from the cloud was incorporated into the rocks of the Earth, but at a very low and relatively uniform concentration.

The initial enrichment of uranium occurred when the majority of the Earth's iron and nickel sank downward to form the core, leaving lighter silicon and oxygen behind to form the mantle. Uranium readily forms chemical compounds with oxygen and was therefore buoyed upward into the mantle rocks. A greater enrichment occurred as the even less dense rocks of the crust began to separate from the mantle. Uranium is a rather large atom that does not fit very well into the densely packed crystalline structures found in the mantle. It therefore tends to be forced upward into the crust as a residual element when the mantle partially melts, a process that is ongoing to this day.

Similar processes of partial melting and upward migration of uranium also occur within the crust itself, so that it has become disproportionately enriched in the uppermost 10–15 km of the continents. The average concentration there is around 3 parts per million (0.0003 percent), about 150 times its planetary average. The remaining natural enrichment required to produce commercial ore, which contains about 0.1 percent uranium at a minimum, is provided by water flowing through the subsurface.

Uranium from Groundwater

Uranium shares an important characteristic in common with fish: It is much more mobile in well-oxygenated waters than in water from which oxygen has been depleted. Oxygen-rich groundwater therefore dissolves uranium from the rocks through which it flows. Uranium is precipitated in solid form again in places where the oxygen content drops and forms highly enriched commercial deposits. A variety of uranium minerals may be naturally precipitated, the most common of which is uraninite (UO_2 with variable amounts of U_3O_8). Because the concentration of uranium in water is small, the precipitation rate is slow, but over long stretches of geologic time a very large deposit can accumulate. You might think of this process as being similar to a municipal water supply that flows through a filter before entering the water main. The water may appear to be quite clear, but over long stretches of time the filter will eventually become clogged with fine deposits and require replacement.

About one-fifth of the world's uranium reserves formed within permeable sandstone beds, which originally served as conduits for oxygenated groundwater to flow downward from the Earth's surface. Eventually these waters encountered realms where the oxygen had been largely exhausted by aerobic decay of dead organic matter, which caused uranium to precipitate. Such deposits are particularly common in the western United States, Kazakhstan, and Australia. Uranium can also be precipitated around isolated bits of organic matter contained in sandstone aquifers, within organic-rich mudstone, or even at the contact between oxygenated waters and petroleum.

At a larger scale, uranium is found concentrated near the lower bounding surface of some sedimentary basins. Two such basins, the Athabasca Basin in northwest Canada and the McArthur Basin in northern Australia, together account for about one-third of the world's known reserves. The Athabasca Basin shares its name with the heavy and extra-heavy oils found in Alberta (see Chapter 11) but lies mostly within Saskatchewan and formed much earlier, approximately 1.5 billion years ago. Uranium is concentrated near its base, where sedimentary layers rest upon underlying igneous and metamorphic rocks. This basal unconformity marks a major shift in past surface conditions, from erosion of the underlying rocks to deposition of overlying sedimentary layers. Like the Cretaceous unconformity that helped to concentrate oil in Alberta, this much older unconformity helped to concentrate uranium. Uranium precipitation appears to have occurred where oxygenated uranium-bearing waters circulating downward through the basin met with hot, oxygen-poor waters rising from below. The richest deposits are found where major faults, which act as fluid conduits, intersect the unconformity.

Moving water gives rise to a variety of other enriched uranium ores as well. For example, the largest deposits in Australia, which hold the world's largest uranium reserves, are found in intensely fractured rocks called breccia, through which massive amounts of water have flowed. Some of the strangest uranium ores, found in northern Ontario and in South Africa, contain uranium minerals that were deposited as river-borne sand grains. Gold miners would immediately recognize this as "placer" uranium, similar in origin to the placer gold that they discover by "panning" river gravels. No modern river carries significant amounts of placer uranium, however, because oxygen in the river water quickly dissolves it. The Canadian and African deposits formed more than 2 billion years ago, when the Earth's atmosphere contained much less oxygen.

Commercial ores range from 0.1 percent to more than 20 percent uranium by weight, but even in the latter case extensive processing is still needed to obtain fuel-grade uranium. Originally this was done by physically removing the ore from surface pits or underground mines and crushing it, then leaching with sulfuric acid. In some deposits the mining step can be omitted, and the uranium is leached directly from "in situ" ore underground. The leachate is chemically processed to produce a mixture of uranium oxides called yellowcake, named for its typically yellow to dark brown color. This is one cake you most definitely do not want to eat; with an average chemical formula of approximately U_3O_8, yellowcake contains about 85 percent uranium by weight.

Like coal mining, uranium mining presents some inescapable environmental consequences, particularly related to mine tailings. Tailings consist of finely ground rock that still contains some uranium, along with other radioactive wastes such as radium-226 (which decays into radon gas) and thorium-230. The tailings may also contain nonradiological contaminants such as heavy metals or arsenic, and sulfide minerals that can form sulfuric acid if exposed to water and oxygen. Acid mine drainage can kill fish and other organisms living downstream, and sulfuric acid can also leach residual uranium from the tailings. Uranium tailings therefore require active management long after the original uranium resource has been depleted, with special attention to preventing erosion or water seepage.

Uranium from the Sea

The ocean represents the largest reservoir of oxygenated water on Earth, and might therefore be expected to contain a large amount of uranium in solution. Estimates of its uranium concentration vary, but generally fall in the range of only a few parts per billion. However, because of the immense volume of the ocean this small concentration adds up. At 3 parts per billion the ocean would contain more than 4 trillion kg total of uranium, which is about one thousand times more than the mass of uranium in all known ores. At first glance it would appear that the oceanic store of uranium could satisfy our energy needs for millennia to come!

The trick, however, is to get the uranium out of the seawater. The Earth actually solved this problem on its own hundreds of millions of years ago, by trapping uranium in phosphorite deposits, organic-rich shale, and coal that was deposited under low-oxygen conditions. In some cases the uranium concentration of such rocks rises high enough to make them viable commercial ores. Uranium in coal can also become a problem, if released during combustion. For example, a well-publicized 1978 study by researchers from the Oak Ridge National Laboratory concluded that the radiation dose released from a coal-fired power plant is actually greater than that associated with an equivalent nuclear facility. The source of this radiation is uranium and thorium contained in "fly ash" that escapes through the smokestack.

No practical method yet exists for humans to extract uranium directly from seawater. To do so would be no small task. Obtaining 1 kg of ^{235}U would require filtering through about 18,000 Olympic-sized swimming pools filled with seawater, assuming 100 percent recovery. A variety of experimental uranium-absorbent materials have been developed in an attempt to accomplish this task, but none has yet reached the point of economic feasibility.

The 4 Percent Solution: Enrichment of Fissile Uranium

As difficult as it is to obtain relatively pure uranium, it is still not ready for the reactor at this point. Fissile ^{235}U makes up only 0.72 percent of the natural isotope abundance, with nonfissile ^{238}U constituting most of the remainder. The reason for this imbalance is simple: ^{235}U, with a half-life of about 704 million years, decays much faster than ^{238}U, with a half-life of 4.5 billion years. Since the Earth first formed, its ^{235}U endowment has declined to about 1.2 percent of its starting mass, whereas 50 percent of the original ^{238}U remains. The relative abundance of these two isotopes has therefore shifted dramatically in favor of ^{238}U.

Current reactor designs require that about 3–4 percent of the fuel consist of ^{235}U, which requires enrichment by a factor of nearly six times relative to natural abundance. Various techniques have been developed to do this, all of which are based on repetitive concentration of ^{235}U based on its slightly lower density. The most commonly used method is to convert the yellowcake into uranium hexafluoride gas (UF_6), which is then placed in a centrifuge. The spinning motion of a centrifuge works something like a fast-moving playground carousel, which requires its riders to hold on tight or risk being thrown off. The centrifuge throws heavier, ^{238}U-bearing gas toward its outer walls, where it can be withdrawn, enriching the remaining gas in ^{235}U. Because the density difference involved is small, it requires many centrifuges connected in a cascade arrangement to produce the desired enrichment. The enriched ^{235}U is then converted into solid UO_2 fuel assemblies that are loaded into the reactor to be "burned," the industry vernacular for the actual fission process. At most commercial reactors (including all those in the United States), once the concentration of ^{235}U has fallen below about 1 percent, the fuel assemblies are removed, and from that point forward they are treated as nuclear waste.

Conclusions

Geothermal and nuclear energy are in many ways two sides of the same coin. Both derive from the primordial Earth's original endowment of energy and therefore are inherently nonrenewable. Both rely either partly or entirely on the release of nuclear potential energy that was originally captured during stellar explosions. Water is used to transfer the heat produced in both cases, and groundwater flow serves naturally to enrich ore deposits of uranium. Neither geothermal nor nuclear energy directly emits carbon dioxide or other greenhouse gases to the atmosphere.

Geothermal energy has the advantage of not directly producing wastes at all, and of potentially being available anywhere on Earth. However, the natural flux of heat upward from the interior of the Earth is only a tiny fraction of the downward flux of sunlight and by itself is far too small to support any useful energy application. Geothermal systems instead must rely on the heat that has been previously stored in rocks. Once it is extracted, replenishment of this heat will occur only very slowly; geothermal energy therefore is not truly renewable over timescales of practical interest. Like fossil fuels, dormant volcanoes tend to be highly localized. Lower-quality geothermal resources are much more widespread, but are also more expensive to exploit.

Nuclear fission in contrast can produce power at rates that are limited only by our willingness to invest in nuclear reactors and their supporting systems. Uranium is among the least abundant elements in the universe, but the geologic evolution of the Earth has fortuitously enriched it in the crust at concentrations millions of times greater than its average planetary abundance. However, most naturally occurring uranium consists of ^{238}U, which cannot support fission in current-generation nuclear reactors that employ "slow" (thermal) neutrons. Further enrichment of ^{235}U, a fissile isotope that naturally constitutes 0.72 percent of naturally occurring uranium, is therefore required.

Although known uranium resources in general appear sufficient to fulfill current needs for thousands of years, ^{235}U resources are much more limited, and conventional ore bodies could by some estimates become exhausted within a century. The ocean contains a much larger resource, but the practicality of extracting it remains in doubt. Nuclear fission also has some other well-known disadvantages, including the risk of accidental radiation release from power plants, the potential proliferation of the technology used for making nuclear weapons, and the production of waste products that will remain radioactive for millennia.

For More Information

Brown, G., and Garnish, J., 2004, *Geothermal energy: An overview*, in Boyle, G., ed., Renewable Energy, 2nd ed.: Oxford, Oxford University Press, p. 342–382.

Burchfield, J. D., 1975, *Lord Kelvin and the Age of the Earth*: New York, Sci Hist Pub, 260 p.

Carbon, M. W., 2006, *Nuclear Power: Villain or Victim?* 2nd ed.: Madison, WI, Pebble Beach, 108 p.

Dahlkamp, F. J., 1993, *Uranium Ore Deposits*: Berlin, Springer-Verlag, 460 p.

Dickson, M. H., and Fanelli, M., eds., 2003, *Geothermal Energy: Utilization and Technology*: Paris, United Nations Educational, Scientific, and Cultural Organization, 205 p.

Duffield, W. A., and Sass, J. H., 2003, Geothermal-clean power from the Earth's heat: *U.S. Geological Survey Circular* 1249, 36 p.

Jaupart, C., and Mareschal, J.-C., 2011, *Heat Generation and Transport in the Earth*: Cambridge, Cambridge University Press, 464 p.

Jefferson, C. W., Thomas, D. J., Gandhi, S. S., Ramaekers, P., Delaney, G., Brisbin, D., Cutts, C., Quirt, D., Portella, P., and Olson, R. A., 2007, Unconformity associated uranium deposits of the Athabasca Basin, Saskatchewan and Alberta, *in* Goodfellow, W. D., ed., *Mineral Deposits of Canada: A Synthesis of Major Deposit-Types, District Metallogeny, the Evolution of Geological Provinces, and Exploration Methods*: Geological Association of Canada, Mineral Deposits Division, Special Publication No. 5, p. 273–305.

KamLAND Collaboration, 2011, Partial radiogenic heat model for Earth revealed by geoneutrino measurements: *Nature Geoscience*, v. 4, p. 647–651.

McBride, J.P., Moore, R. E., Witherspoon, J. P., and Blanco, R. E., 1978, Radiological impact of airborne effluents of coal and nuclear plants: *Science*, v. 202, p. 1045–1050.

McKay, A. D., and Miezitis, Y., 2001, Australia's uranium resources, geology and development of deposits: AGSO - Geoscience Australia, Mineral Resource Report 1, 196 p.

Murray, R. L., *Nuclear Energy - an Introduction to the Concepts, Systems, and Applications of Nuclear Processes*, 4th ed.: Oxford, Pergamon Press, 437 p.

Pollack, H. N., Hurter, S. J., and Johnson, J. R., 1993, Heat flow from the Earth's interior: Analysis of the global data set: *Reviews of Geophysics*, v. 31., p. 267–280.

Pyle, D. M., 1995, Mass and energy budgets of explosive volcanic eruptions: *Geophysical Research Letters*, v. 22, p. 563–566.

Taylor, S. R., and McLennan, S. M., 2009, *Planetary Crusts*: Cambridge, Cambridge University Press, 378 p.

Tester, J. W. (and 17 coauthors), 2006, *The Future of Geothermal Energy: Impact of Enhanced Geothermal Systems (EGS) on the United States in the 21st Century*: Cambridge, MA, Massachusetts Institute of Technology, 372 p.

14

Out of Sight, Out of Mind: Geologic Waste Disposal

All places are alike,
And every earth is fit for burial.

Christopher Marlowe, Edward II *(1594)*

There is something deeply satisfying about burial in the Earth. On the one hand, it can symbolize hope for the future, for example, if we sow the seeds for a new crop or bury something of value that we plan to dig up later (treasure perhaps, or a time capsule). On the other hand, burial is wonderfully effective for concealing that which we wish to permanently forget, such as the inevitable decay of our deceased relatives or the mountains of household trash we produce every day. The former we reverently inter in cemeteries; the latter we dump unceremoniously into landfills. The net result is the same in both cases, however; a problem has been eliminated and we can move on with our lives.

It is tempting to believe that energy wastes can be eliminated the same way, and perhaps they can. Saline water (brine) that has been produced together with crude oil or natural gas has been routinely reinjected into deep disposal wells for decades. If released untreated at the surface, subsurface brines can harm vegetation, wildlife, and livestock and contaminate freshwater supplies. Returning them to their point of origin provides an expedient, low-cost solution that generally causes little trouble. In a relatively small percentage of cases this solution may create new problems of its own, however, if high-pressure fluid injection unlocks previously stable earthquake faults.

Burial might also be used to dispose of unwanted CO_2, to limit its buildup in the atmosphere. While not yet routine, this approach is currently being tested at various sites worldwide. Ironically, one of the main challenges is obtaining large volumes of concentrated CO_2

to bury. Its concentration in the atmosphere is very small, currently about 400 parts per million. We would like to bury *just* CO_2 and leave the rest of the atmosphere where it is, but separating the two would be very expensive. Most current research therefore focuses on capturing CO_2 emitted from coal-fired power plants or other point sources, before it can mix with the atmosphere. In this concentrated form CO_2 can be easily and efficiently pumped underground. Burying newly generated CO_2 has no effect on previous emissions that have already entered the atmosphere, but it could reduce future emissions and buy time for existing greenhouses gases to decline naturally.

Finally, burial may be the preferred means of disposing of spent nuclear fuel. This material is precisely the opposite of a dilute atmospheric gas; it is a concentrated and very dense solid that can be loaded onto a truck or railcar. It might therefore seem a simple matter just to stack it up in abandoned mine tunnels or bury it in open pits, and call the job done. Unfortunately the long lifetime of radioactive isotopes contained in spent fuel precludes such a simple solution. Whereas dead relatives and household trash can both be expected to decay in benign obscurity, the fission products in spent fuel will continue to "glow in the dark" for more than a century, and other components will emit radiation for many thousands of years.

It is not considered sufficient just to make nuclear waste disappear from sight. We also want assurances that it will stay put, protected from leakage, accidental exhumation, or unauthorized retrieval, over periods that exceed the entire duration of recorded human history. This is no easy feat. The last society to try something so ambitious may have been ancient Egypt, whose people built massive pyramids and underground tombs to guarantee their pharaohs and other elites an undisturbed afterlife. Those efforts proved largely futile, however; looters began to penetrate the tombs almost as soon as they were sealed. It remains to be seen whether our modern society can do any better.

Faults and Fluids: Induced Seismicity

The 1985 James Bond film *A View to a Kill* featured an evil industrialist who plans to destroy the Silicon Valley area of California by flooding the Hayward and San Andreas Faults with lake water, thereby inducing a catastrophic earthquake. Fortunately Bond thwarts the plot, then finishes off the villains in an exciting action sequence involving a blimp, some dynamite, and the Golden Gate Bridge.

While the movie plot was fanciful, it contained a kernel of truth. Water and other fluids can indeed induce movement on earthquake faults if they are pumped into a fault zone under pressure. Generations of geology students have been taught this principle using an empty beer can (Figure 14.1)! The beer can is placed inverted on a glass plate, which is then tilted upward until the can overcomes friction and begins to move. This experiment is meant to simulate the natural movement of a fault. Under normal conditions the plate must be tilted to an angle of approximately 17° or greater before the can begins to slide.

Next, a second empty beer can that has been chilled in a freezer is removed and quickly placed upside-down on the plate. This time the can begins to slide at a much smaller tilt angle, as little as 1°. It does so because the air inside the can is warming up, and as it does so, its pressure increases. This in turn reduces the friction between can and the glass plate, simulating a fault zone that has had fluid pumped into it. Readers are invited to try this experiment themselves; please note that non-alcoholic soft drink cans are reported to produce a similar result.

The effects of fluid injection into faults are not merely academic, but have been demonstrated many times. For example, between 1962 and 1965 the U.S. Army injected contaminated water from the Rocky Mountain Arsenal weapons plant, near Denver, Colorado, into a disposal well drilled to a total depth of 3,671 m. Local seismographs began to register earthquakes in the area less than two months after injection began. Most were too small to notice, but some reached magnitudes of about 4, enough to cause minor damage. Over the next several years the frequency of earthquakes correlated closely with the volume of fluids injected into the well (Figure 14.2), and injection was finally halted in 1965. The earthquakes continued for some time afterward, however, with three more damaging magnitude 5 events recorded in 1967.

Similar experiences have been repeated many times since and are not limited to wastewater injection. Geothermal energy projects have induced perhaps the greatest number of earthquakes, averaging 300–400 per year at The Geysers in California (see Chapter 13). In 2006, Basel, Switzerland, experienced several small but noticeable earthquakes associated with water injection for a deep geothermal energy project. The project was discontinued out of fear of triggering a larger quake, such as the one that leveled the city in 1356.

Induced earthquakes have also been known to occur in areas of active oil and gas development. Some of these resulted not from injection but from extraction of hydrocarbon fluids, leading to partial

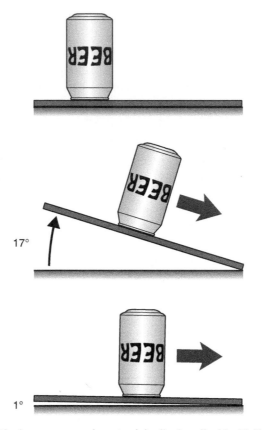

Figure 14.1. The beer can experiment originally described by M. K. Hubbert and W. W. Rubey in 1959 to help explain movement on low-angle faults.

Top: An emptied room temperature beer can is placed upside-down in a puddle of water on a sheet of glass. Middle: One edge of the glass is tilted up until the can begins to slide, and the angle of tilt noted. Bottom: The experiment is repeated with a can that has first been chilled in a freezer. This time, the can slides at a much lower tilt angle because of the pressure of warming air inside. The increased pressure supports some of the weight of the can, reducing the friction where its rim contacts the glass.

collapse of the reservoir. As noted in Chapter 10, earthquakes asso-ciated with hydraulic fracturing are generally too small to be felt at the surface and are not considered threatening. More noticeable earthquakes can result from injection of wastewater from hydraulic fracturing into disposal wells, however. For example, a series of small

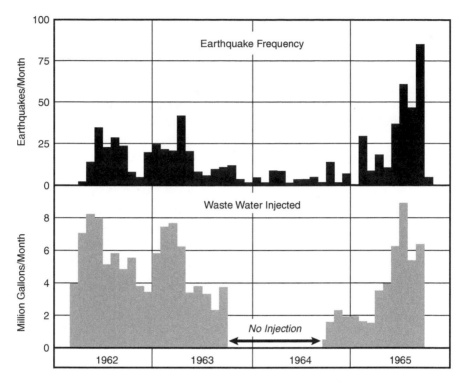

Figure 14.2. Seismicity related to wastewater injection at the Rocky Mountain Arsenal, Denver, Colorado (redrawn using data reported by Healy et al., 1968).

(magnitude 2.5–3.3) earthquakes occurred near a brine injection well located at the Dallas–Fort Worth Airport in 2008–2009. No seismicity had been recorded in this area prior to that time.

Despite these occurrences, the vast majority of water injection wells (which exceed 100,000 in the United States) have been operated without any problems at all, in some cases for decades. The potential for trouble appears not to be universal, but instead to be linked to the specific geology of the injection site and to the manner in which injection is accomplished.

Putting the Genie Back into the Lamp: CO₂ Storage in Petroleum Reservoirs

The mythical figure Aladdin is famous for summoning a genie from inside an oil lamp, who then grants him untold riches. This tale might be viewed as a parable for the technological advances enabled by petroleum during the 20th century, which would have appeared almost magical to those living during earlier times. Oil and gas were extracted from the ground rather than a lamp, but the resultant enrichment of human lifestyles has been no less impressive. However, there has also been an unintended dark side to this magic: massive release of CO_2 and the resultant alteration of Earth's atmospheric heat balance.

What if we could keep the riches but return the CO_2 to its source? This idea is not so far-fetched as it might seem; the underground reservoirs from which oil and gas have been extracted also represent convenient repositories for reinjected CO_2. Most of the investment required to map out these reservoirs has already been made, and some of the existing infrastructure can be reused. Oil companies have already developed tried-and-true technology for injecting CO_2 into the ground and have used it to boost oil production from many older fields (Figure 14.3).

The petroleum industry currently gets most of its CO_2 from natural underground accumulations. Unfortunately, reinjection of such CO_2 does not provide any net reduction in greenhouse gases; it merely transfers CO_2 from one underground reservoir into another. Obtaining and transporting this CO_2 can also be surprisingly costly. The National Energy Technology Laboratory estimated in a 2010 report that typical CO_2 purchase prices were in the range of $10.00–$15.00 per metric ton, and that the total costs including transportation and injection were around $38.50 per ton. They concluded that these costs can represent anywhere from 25 percent to 50 percent of the cost of a barrel of oil obtained in this fashion.

The same job could be done using waste CO_2 obtained from the combustion of fossil fuels or other human activities. A distinct advantage of this approach is that the sale of produced oil can offset some of the cost of collecting, transporting, and injecting the waste CO_2. The most prolific source of concentrated waste CO_2 is power plants burning fossil fuels, which according to the International Energy Agency are responsible for about 40 percent of human-induced CO_2 emissions globally. Other point sources include oil refineries and ethanol refineries. A surprising 5 percent of global emissions originate from cement plants. Cement plants start with limestone ($CaCO_3$) as a raw ingredient and then roast it at 600°C–900°C to make lime (CaO), the main ingredient in cement. About half of the CO_2 emission results from fuel combustion to heat the limestone, and the other half emanates from the limestone itself. The technology for capturing CO_2 from

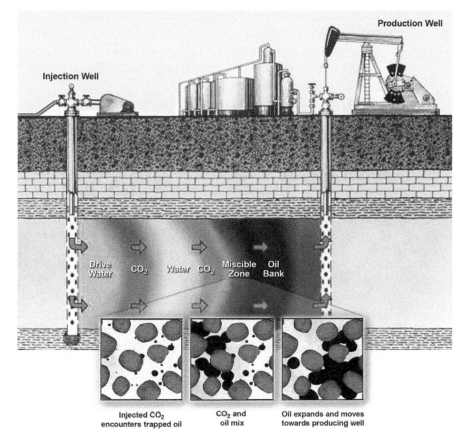

Figure 14.3. Schematic illustration of the use of CO_2 injection to enhance crude oil production (National Energy Technology Laboratory).

point sources already exists, although doing so will not be cheap. The Intergovernmental Panel on Climate Change (IPCC) estimated in 2005 that CO_2 capture and compression alone would add between 20 percent and 70 percent to the cost of electrical generation.

Numerous pilot projects are currently exploring the potential for storing CO_2 in depleted oil fields; one of the largest and best known of these is located in southern Saskatchewan. In 1954 a moderately large oil field was discovered near the small plains town of Weyburn, which had previously been known principally as a Canadian Pacific Railway stop and the site of a major mental health facility. By the late 1990s oil production was declining, despite attempts to prop it up with additional drilling. Since 2000, however, the field has experienced a rebirth, due to injection of CO_2 obtained via pipeline from a coal gasification plant across the border in North Dakota. It is expected to produce oil for another 20 to 25 years, and the National Energy Technology Laboratory has estimated that some 30 million metric tons of CO_2 will ultimately be stored there.

This last statistic unfortunately highlights a fundamental limitation of combining CO_2 storage with enhanced oil recovery: The total volume of oil reservoirs is simply too small to have a significant effect on global warming. Furthermore, not all of the volume that does exist can be used. The utility of CO_2 injection in boosting oil recovery varies widely from field to field, and many oil fields lack sufficient integrity to prevent rapid leakage of CO_2 back to the surface. In 2012 the U.S. National Energy Technology Laboratory estimated that something between 6.5 and 45 billion tons of CO_2 could be stored in U.S. oil reservoirs. This translates into only one to eight years of storage capacity at current rates of U.S. CO_2 emission.

A larger reservoir volume is available in natural gas reservoirs. These offer less opportunity for offsetting the costs of CO_2 injection, however, and would themselves likely be filled within a matter of decades. Coal might offer some additional help, via the injection of CO_2 into coal seams that cannot be mined. Once injected, CO_2 sticks onto surfaces of organic particles in the coal. The U.S. 2012 *Carbon Utilization and Storage Atlas* estimated that ultimate storage capacity at only about one-quarter to one-half that in depleted oil and gas fields, however. In the end, the genie simply won't fit back into the lamp.

The Salty Solution: CO_2 Storage in Saline Aquifers

Conventional oil and gas reservoirs are by definition geologic anomalies, constituting only a tiny fraction of the total volume of potential underground storage space. Most of the available pore space in rocks is filled with water, most of which is salty. These saline aquifers could potentially store large amounts of CO_2; for example, the U.S. 2012 *Carbon Utilization and Storage Atlas* estimated their capacity to be between 10 and 100 times greater than that of depleted oil and gas fields. Figures for other countries are likely to reveal similar relationships. Even the lowest capacity estimates for saline aquifers are large enough to provide at least several centuries worth of CO_2 storage at present emission rates.

The uncertainty in these estimates in part reflects the fact that saline aquifers are largely unexplored, and therefore relatively poorly delineated. Literally millions of wells have been drilled in the attempt to find or produce oil and gas, whereas very few have been drilled specifically for brine. Brine generally is regarded as a waste product, except in a few locales where it contains high concentrations of bromine, potash, soda ash, or other commercially valuable salts. Most of what we know of the geology of saline aquifers therefore has been learned accidentally, through inadvertent encounters during the drilling for more valuable oil and gas. While these data are useful, oil and gas wells tend to focus on areas of known or suspected traps, which most often are geologic high points such as anticlines. Conversely, most of the volume of saline aquifers lies at lower levels where relatively few wells have been drilled.

This shortage of drilling data puts saline aquifer storage at a severe disadvantage compared to oil and gas fields. For many areas an entirely new exploration drilling program will be required to document the suitability and capacity of saline aquifers for CO_2 injection. The costs for such projects would rival the staggering sums already invested in oil and gas development. Existing saline aquifer injection projects therefore mostly coincide with areas of active oil and gas production, where the underground geology is already well known. One of the largest examples is the Sleipner field in the North Sea, offshore Norway, where CO_2 is separated from produced natural gas and then reinjected into a shallower saline aquifer (Figure 14.4). The project is motivated in part by a carbon emissions tax imposed by the Norwegian government. To date, about 8 million tons of CO_2 has been injected into the aquifer, a drop in the bucket compared to total global emissions but enough to help demonstrate the feasibility of the project.

Saline aquifers need to meet several criteria in order to receive CO_2, starting with a minimum depth of injection of approximately 800 meters. CO_2 injected above this depth remains in a gaseous state (Figure 14.5), which makes it relatively bulky and increases the chance it will quickly bubble back to the surface. At the greater depths CO_2 transitions into a new state, called a supercritical fluid. Supercritical does not imply that the CO_2 has become dangerous or even disagreeable; it just means that distinct phases (solid, liquid, gas) no longer exist. Supercritical CO_2 occupies far less volume, making it much easier to store.

CO_2 storage must also not contaminate overlying freshwater aquifers. Of course, drinking small quantities of dissolved CO_2 is not necessarily a problem; in fact, the multibillion dollar soft drink industry intentionally adds CO_2 to its sweetened beverages. The effect is

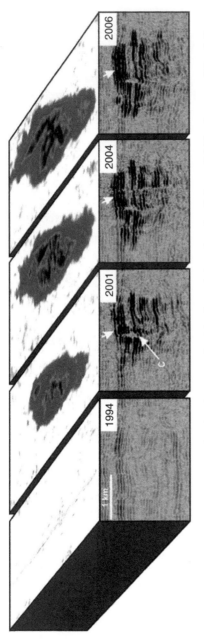

Figure 14.4. Images showing progressive development of a CO_2 plume injected at the Sleipner field in the central North Sea between the United Kingdom and Norway.

Injection commenced in 1996, and darker tones generally indicate increasing CO_2 content through time. Front panels show the response of sound (seismic) waves reflecting off porous rock layers that contain CO_2, as viewed from the side. Top panels show the plume and intensity of seismic response, as viewed from above (note that the scales of side and top views are not equal). The area labeled "C" indicates the main CO_2 feeder channel, and the white arrow pointing downward indicates the approximate top of injected CO_2 (modified from Chadwick and Noy, 2010).

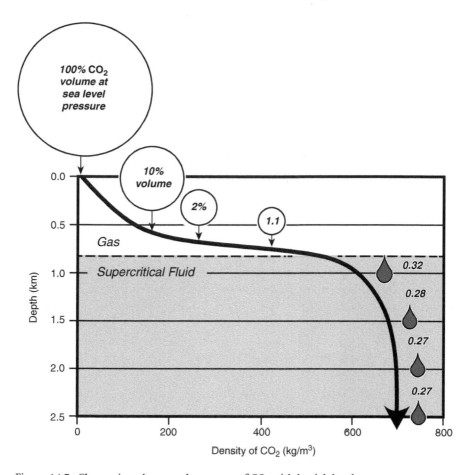

Figure 14.5. Change in volume and pressure of CO_2 with burial depth.
The volume occupied by a given mass of CO_2 gas decreases rapidly with burial up to about 800 m, as a result of increasing pressure. Below 800 m the CO_2 becomes a supercritical fluid, which changes little in volume at greater depths (Modified from National Energy Technology Laboratory, 2012).

generally described as delightful. It is doubtful, however, that most people would want their domestic water supply to consist exclusively of the equivalent of club soda or Perrier™ water. At some point the novelty would wear off, and the acidic nature of carbonated water would take a toll on tooth enamel and household plumbing. Fortunately, nearly all domestic water supplies lie above the 800 m level that marks the upper depth limit of planned CO_2 injection.

Leakage of stored CO_2 back to the atmosphere could in extreme cases pose a safety risk of its own. For example, a natural release of volcanic-sourced CO_2 from Lake Nyos in Cameroon killed an estimated seventeen hundred people. The CO_2 had built up gradually over time in the bottom waters of the lake, and then suddenly rose through the water column without warning in 1986. Although the probability of such catastrophic release is exceedingly small, this event graphically illustrated that CO_2 is not an entirely benign substance.

Like oil and natural gas, supercritical CO_2 is less dense than salt-water and therefore tends to rise upward through water-filled pore spaces. If not blocked by impermeable layers such as shale or salt, it might therefore seep all the way back to the surface. The way CO_2 interacts with water tends to slow this upward seepage, however. As a plume of supercritical CO_2 rises through the subsurface it leaves a mixture of water and small blobs of supercritical CO_2 in its wake (Figure 14.6). These blobs tend to remain immobilized within individual pores, because of the capillary pressures exerted by water. Unlike oil, CO_2 is readily soluble in water, and given enough time the blobs will dissolve to form carbonic acid (H_2CO_3). This is the same basic reaction used to make carbonated drinks, and once in solution CO_2 is only as mobile as the water body itself. Because deep aquifers typically have very slow underground flow rates, the dissolved CO_2 should remain underground for thousands of years.

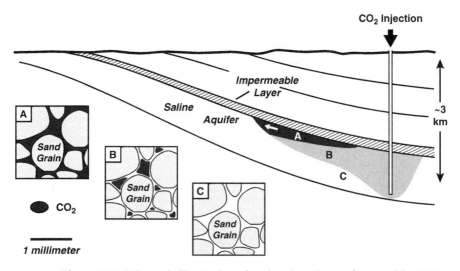

Figure 14.6. Schematic illustration of a migrating plume of supercritical CO_2 that has been injected into an aquifer.

Turning CO$_2$ into Stone

The most permanent solution to CO$_2$ storage would be to convert it into solid rock. Given enough time, the Earth can actually do this job for us. Carbonic acid in saline aquifers eventually reacts with its host rocks via the same kinds of chemical reactions that dissolve rocks exposed on the surface of the continents (see Figure 2.6). Once released they can form new solid compounds with carbon, which suitably enough are known as carbonates. In most natural settings this happens rather slowly, however; it would likely take centuries or millennia fully to convert injected CO$_2$ into carbonate rocks.

We can do it faster, but only at a price. Rather than waiting for carbonates to form slowly within aquifers, we can mine the raw materials needed and combine them industrially with CO$_2$ obtained from point sources. One of the best materials to use is olivine, a silicate mineral named for its olive-green translucent color. Gem-quality olivine is also known by the name peridot, and rocks that are made mostly of olivine are therefore called peridotite. Peridotite rocks are native to the mantle, hundreds of kilometers underground, but can rise to the surface at midocean ridges or be shoved upward during the violent collisions of moving tectonic plates.

Olivine is also one of the most chemically reactive minerals known. Given a chance, the magnesium in olivine will eagerly combine with CO$_2$ to make a carbonate compound called magnesite (MgCO$_3$). Reaction rates can actually be quite fast, on the order of hours, but only if the peridotite is first ground into a powder, the pressure of CO$_2$ is raised to about one hundred times atmospheric pressure, and the whole experiment is preheated to about 185°C.

All of this mining, transport, grinding, compressing, and heating is costly in terms of the capital, labor, and energy required. The mining step alone is daunting; in 2005 the Intergovernmental Panel on Climate Change estimated that between 1.6 and 3.7 tons of rock would be needed to react with 1 ton of CO$_2$. To put this in perspective, a typical large family automobile produces between 5 and 10 tons of CO$_2$ per year. At these rates, anywhere from 8 to 37 tons of peridotite would therefore need to be mined every year to erase the emissions of a single vehicle that weighs about 2 tons. Add to this the need to dispose of large amounts of magnesite, a brand new waste product created in the process, and the whole idea begins to look like bad judgment.

If we could somehow introduce CO$_2$ directly into peridotite rock bodies we could take advantage of olivine's chemical reactivity without the need to dig it up. Geologist Peter Kelemen at Columbia University and colleagues recently reviewed two possible methods for doing just this. The first method involves drilling wells into areas of peridotite rock, heating the rocks up to the requisite temperature, hydraulically fracturing them, and then injecting high-pressure CO_2. This method shares a lot in common with the enhanced geothermal systems discussed in Chapter 13, except that rather than extracting heat from the rocks, heat would initially need to be added.

The second method involves drilling offshore wells into mid-ocean ridges. Seawater naturally circulates through these areas and carries with it dissolved CO_2, which is precipitated as carbonate minerals. This natural process could in theory be accelerated by hydraulically fracturing the rocks beneath midocean ridges. While elegant in principle, however, such a scheme would be staggeringly expensive in application. Kelemen and colleagues estimated that 1 *million* wells would be needed to dispose of 1 billion tons of CO_2 per year, a small fraction of current global emissions. All of the world's current supply of oil and natural gas is produced from about the same number of wells, and most of those wells were drilled onshore at much lower cost. The idea of effectively reproducing the entire global petroleum industry in deep offshore waters is difficult to fathom.

Burial at Sea

Humans have long dumped various wastes into the sea, presumably with the idea that the sea is big and therefore no one will notice. As it turns out we are already doing the same with CO_2, although not intentionally so. CO_2 continually moves back and forth between the atmosphere and the ocean, alternately dissolving into and degassing out of the water. Over time these fluxes tend toward a kind of balance, which is dominated by the ocean because it contains much more CO_2.

The recent increase in CO_2 emissions has tilted the balance, so that CO_2 is now entering the ocean faster than it is leaving. Eventually the ocean should be able to absorb most of the new emissions, but unfortunately it can't do so fast enough to keep atmospheric concentrations stable. Even though the capacity of the ocean is huge, its rate of absorption is limited by how fast CO_2 can be transferred across its surface. In 2005 the Intergovernmental Panel on Climate Change estimated that the ocean was absorbing only about one-quarter of the new emissions.

We could speed up the transfer by injecting concentrated CO_2 directly into deeper waters. If done at shallow depths this would not help much because liquid CO_2 would quickly rise to the surface and bubble back into the atmosphere. Below about 3,000 meters depth liquid CO_2 becomes more dense than seawater, though, and it would settle to form submarine lakes on the ocean floor. The engineering required to extend pipes into deep water is well established within the petroleum industry, and the cost of this approach would presumably be far lower than the peridotite carbonation schemes described previously.

Unfortunately this kind of storage only kicks the problem into the future; the IPCC estimated that some of the injected CO_2 would begin to find its way back to the atmosphere within a couple of centuries. After a millennium or two the net result would be about the same as if nothing had been done at all. The creation of vast submarine lakes of CO_2 would also introduce a host of new environmental impacts, starting with the asphyxiation of whatever organisms might happen to be living in the area at the time. The practical consequences are largely unknown, but seem unlikely to be good.

The Gift That Keeps on Giving: Spent Nuclear Fuel

It would be fair to say that the problem of CO_2 emissions sneaked up on us quietly; after all, humans happily burned coal for centuries before anyone began to suspect that fossil fuel combustion might actually alter global climate. Only within the past fifty years or so has this awareness fully dawned within the scientific community, and some among the public still have not accepted it. In contrast, the potentially destructive capabilities of nuclear energy literally announced themselves with a bang and became instantly impossible to ignore. The detonation of nuclear weapons instilled in many an enduring sense of fear, and it seems undeniable that fear has also to some extent colored public perceptions of nuclear power generation. Incidents such as those at Three Mile Island, Chernobyl, and Fukushima have only reinforced this feeling.

In addition to being intrinsically more frightening, the leftovers of nuclear power generation are also considerably more complex than those associated with fossil fuel combustion. Rather than focusing our attention on a single, simple compound that is exhaled by every human (CO_2), we must confront a bewildering variety of exotic elements and their isotopes, which constantly change their composition

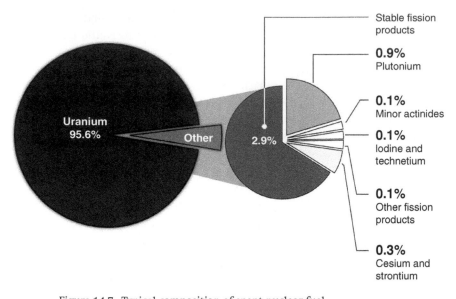

Figure 14.7. Typical composition of spent nuclear fuel.

Note that ~97% consists of potentially reusable material (General Accounting Office analysis of U.S. Department of Energy data).

through invisible processes of radioactive decay. Nothing in our common everyday experience really equips us to comprehend such realities.

Radioactive waste takes many different forms, ranging from clothing that has been contaminated with short-lived radioisotopes used for medical or industrial purposes to the spent fuel removed from nuclear reactors. None of these can be considered very palatable, but the latter is perhaps the most distasteful because of its large volume and highly radioactive nature. As noted in the previous chapter, the spent fuel removed from most current generation nuclear reactors consists mostly of ^{238}U (Figure 14.7).

If spent fuel contained only uranium, then its handling might be a little more straightforward. For example, it could in principle be mixed back into the crushed rocks from which it was separated (called mine tailings) and reburied in mines. After all, the original uranium ore was peacefully minding its own business before someone dug it up, and in most cases it presented no particular threat. However, putting the nuclear genie back into the bottle is not quite so simple as it might seem. Pulverized tailings have far more reactive surface area than the original ore, and therefore much greater potential to be chemically leached by oxygenated groundwaters. Though the initial radiation level might be lower, the potential for radioactive material to spread beyond the mine would be much greater.

Unfortunately high-level nuclear waste contains more than just uranium; it also includes plethora of new isotopes that were never present in the original fuel. These newcomers fall into two main camps. The first includes isotopes with atomic masses relatively close to that of uranium, referred to as actinides because they contain at least the same number of protons as the chemical element actinium (atomic number = 89). Uranium, with 92 protons, is included in this group. So too are elements such as neptunium (Np) and plutonium (Pu). The actinides are all radioactive and some, such as ^{239}Pu, are also fissile. They can therefore potentially be used as reactor fuel or bomb material.

The second group of new isotopes, called "fission products," result from the splitting of uranium into smaller atoms. Fission products are also radioactive, with half-lives ranging from less than 30 years up to 15.7 million years. Those with very short half-lives can pose a major hazard over relatively short time frames of tens to hundreds of years. In some cases this radioactivity can be useful; for example, ^{137}Cs (cesium) is used as a radiation source for medical and industrial applications. On the other hand, ^{137}Cs was also the principal source of radiation associated with the Chernobyl accident, and its toxic effects are not to be taken lightly.

Because of its diverse isotopic composition, the potential hazards posed by spent nuclear fuel change continuously through time. When first removed from the reactor, the fuel assemblies are highly radioactive and hot, mostly because of short-lived fission products. Without continuous cooling the spent fuel literally would remain too hot to handle or even safely store for the better part of a century. As time goes by these short-lived isotopes decay, and after a century or so longer-lived actinides take over as the dominant source of radiation. Storage systems for spent nuclear fuel must therefore meet a dual challenge: They must safely isolate intensely radioactive material over time frames on the order of one hundred years, but also ensure the security of less radioactive material essentially forever.

Eternal Entombment: Deep Geological Repositories

The same basic characteristic that limits the upward flux of geothermal heat also makes underground storage of spent nuclear fuel attractive: Rocks are good insulators. They can therefore provide a highly effective shield against the alpha particles, beta particles, and gamma rays. Underground storage also provides a convenient means of keeping this material away from those who might want to misuse it. It is easy enough to guard the door of an underground cavern, and any attempt to tunnel in would surely be noticed. Perhaps most importantly, geological storage of nuclear waste conveys a sense of permanence that seems well suited to the requirement for continued security into an indefinite future.

Planning for the storage of nuclear wastes began almost simultaneously with the development of the first commercial nuclear power stations. For example, an early discussion of this problem appeared in a 1957 report of the U.S. National Academy of Science, led by Princeton University geologist Harry Hess. The report suggested three principal alternatives: underground storage in salt mines, storage of solid waste bricks "in sheds in arid areas or in dry mines," and injection of liquid wastes into porous underground reservoirs deeper than 5,000 feet. On the basis of these considerations the report concluded that no suitable sites existed along the Eastern Seaboard of the United States, where ironically the density of nuclear reactors is greatest. The report also pointed to the need for thorough geologic investigation of any proposed site, but presciently warned, "Unfortunately such an investigation might take several years and cause embarrassing delays in the issuing of permits for construction."

Other suggestions include storing nuclear wastes within Antarctic ice, dumping it at sea, firing it into space, or burying it in subduction zones where it will eventually be carried back to the mantle. While creative, some of these suggestions unfortunately appear to violate common sense, international law, or both. Most countries with nuclear power capabilities have instead decided to take the more conservative route of storing spent nuclear fuel in tunnels excavated hundreds of meters underground. The proposed designs for such repositories vary, but generally they aim to ensure that radiation greater than the natural background level cannot escape beyond a specified spatial boundary, over a specified period. The last of these requirements may be the hardest to satisfy, since the periods in question exceed the entire duration of recorded human history.

It seems logical therefore to place underground repositories in areas where the risks of disruption by earthquakes or volcanoes are minimal. Tectonically stable regions within continental interiors are therefore preferred. Ideally the wastes should also be stored away from areas of moving groundwater. Salt deposits offer a unique advantage in this regard, because they are by definition dry. Salt has the added advantage of being naturally malleable; over long periods it can flow in a manner similar to toothpaste. This flowage should eventually collapse the salt around engineered waste repositories, sealing them off against any human or natural intrusion.

The possibilities for underground waste storage are not limited to any particular geological age or rock type, however. Germany has planned to deposit its wastes in Permian salt, but France is planning to bury it in Jurassic mudstone, and several other countries plan to store wastes in granite (see the Geologic Timescale, Figure 2.4). The United States has focused its attention on Tertiary volcanic rocks in Nevada, despite the fact that this region is neither seismically nor volcanically quiescent (Figure 14.8). It appears that all of these solutions are potentially workable, however, at least over periods that humans can reasonably be expected to care about. The chief uncertainties are probably those related to future groundwater interactions. The potential for seismic or volcanic disruption can be reasonably deduced from the local geologic histories, but groundwater is influenced by climatic factors that can change more quickly. Presently proposed repositories lie either above the water table in dry regions like Nevada or below the water table at depths where groundwater flow is very slow.

At present none of the proposed geologic repositories mentioned is actually receiving high-level radioactive wastes, because of various political and technical difficulties surrounding their selection or design. The Yucca Mountain facility in southern Nevada could perhaps be considered the poster child for such problems. It was first conceived in the late 1970s, and research and development activities continued there for nearly as long as singer Wayne Newton has performed in Las Vegas. The project was finally suspended in 2010, after the expenditure of some $10 billion (as of this writing the talented Mr. Newton is still going strong).

Some of the delays encountered in developing geologic repositories appear to result directly from conflicting design criteria; for example, it is difficult to build a repository close to the nuclear reactors that produce waste, but also distant from the populations those reactors serve. Nevada is sparsely populated, but its citizens have consistently said "no thanks" to spent nuclear fuel transported from more populous states.

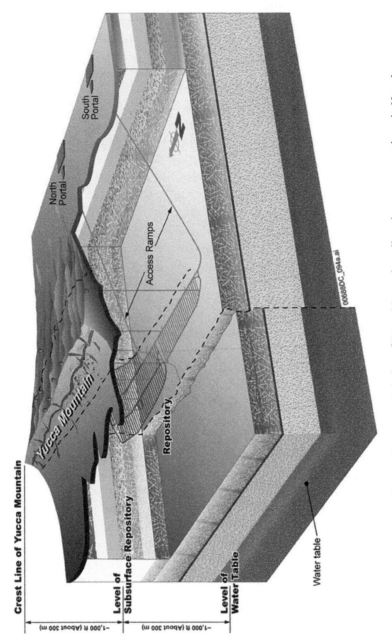

Figure 14.8. Block diagram illustrating proposed design of Yucca Mountain radioactive waste repository in Nevada. The rock layers depicted consist mostly of volcanic tuff, deposited by ash falling to the ground from ancient volcanic eruptions similar to the eruption of Mt. Vesuvius or Mt. St. Helens. These layers are cut by faults (vertical discontinuity between layers), which are believed to be inactive (U.S. Department of Energy).

Other concerns appear to be less realistic. For example, the proposed time frame over which the Yucca Mountain facility would be required to maintain its integrity was initially 10,000 years, but eventually lengthened to 1 million years. Natural environmental changes will surely pose a far greater hazard over such long stretches of the future; for example, the Earth has experienced five major ice ages over just the past 500,000 years. The peak of the last one occurred only about 20,000 years ago, at which time glacial ice sheets up to two miles thick lay across cities from Chicago to Boston. Should such conditions return in the future, leaking radioactive waste may be among the least of our problems.

Fossil Fission and the Long-Term Stability of Radioactive Waste

Predicting the future behavior of deep geological repositories involves an unavoidable element of chance; there are many unknowns and very little experience to learn from. Nuclear fission itself was first achieved only about seventy years ago. Even if it had been discovered by the ancient Egyptians we would not know what might happen to high-level nuclear wastes 10,000 years or more after its removal from a reactor. What we really need is to study the aftermath of a nuclear reactor operating a million or more years ago, but of course none existed then.

Or did they? In 1972 it was discovered that certain uranium deposits in the Oklo region of the African country of Gabon contain anomalously low concentrations of ^{235}U, the fissile isotope that serves as the main fuel for nuclear reactors. ^{235}U normally constitutes 0.72 percent of naturally occurring uranium, with the rest consisting mostly of ^{238}U. At Oklo a small part of the expected ^{235}U appears to have gone missing, however. The deficit cannot be accounted for by normal geochemical processes such as dissolution, because ^{235}U and ^{238}U are virtually identical in their chemical properties. Instead, it appears that some ^{235}U must have been fissioned in a naturally occurring nuclear reaction! Closer examination has yielded definitive evidence that this did in fact occur at sixteen separate sites in the Oklo-Bangombé region, nearly 2 billion years ago.

Such a natural nuclear reactor would be impossible today because the present concentration of ^{235}U in uranium ores is too small to sustain the necessary chain reaction. ^{235}U was much more abundant at the inception of the Earth, but its relative abundance has declined since because it decays much more quickly than ^{238}U. When the Oklo deposits formed about 2 billion years ago, ^{235}U would have constituted about 3.5 percent of the uranium present, a concentration very similar to that of the artificially enriched fuel used in most nuclear reactors today. In addition to having a critical mass of ^{235}U, a natural reactor would also need a neutron moderator in order to operate. Water served that purpose, just as it does in nuclear reactors now. As groundwater seeped into the uranium deposits it slowed down fast-moving neutrons and thereby enabled sustained fission to occur. The heat generated by fission quickly boiled the water away, presumably expelling it to the surface as geysers. The reactor then shut down temporarily until groundwater could seep back in, possibly as soon as a few hours later.

This cycle repeated itself countless times over a period of hundreds of thousands of years, until the remaining natural reactor fuel could no longer sustain fission. Interestingly, the radioactive by-products that have been preserved at Oklo ever since look very much like the spent fuel removed from a modern reactor, if we allow for the effects of 2 billion years of additional natural decay. More to the point, the expected remaining amount of actinides such as plutonium can be mostly accounted for within a few meters of the reaction sites, along with many of the fission products and their descendants. The long-term geologic stability of the site, the presence of impermeable clay layers, and chemically reducing conditions all appear to have played a role in the immobilization of Oklo radioactive wastes. They have remained immobile over a period that exceeds even the most zealous modern repository specifications by a factor of 2,000!

The immobilization of radioactive wastes at Oklo is not perfect; some elements that were either gaseous or soluble under reducing conditions have been lost. On the whole, however, it appears that even accidental geologic storage can be surprisingly effective at sequestering radioactive wastes away from the surface environment. It therefore seems highly probable that modern waste storage systems can be successfully engineered to achieve these same objectives.

Making the Least of It: Fast Reactors and Fuel Reprocessing

The only geological waste repository that can be 100 percent guaranteed never to leak is one that never receives any waste. However, the waste that has already been generated has to go somewhere, and presently it is being stored at hundreds of surface sites worldwide.

As it happens the term "waste" is doubly appropriate, because about 95 percent of this material consists of ^{238}U. While not a fissile reactor fuel itself, ^{238}U is known as a fertile material because it can be converted to fissile ^{239}Pu through the capture of a neutron and subsequent β-decay. This happens routinely during the normal operation of most reactors today, and fission of the resultant ^{239}Pu is responsible for about one-third of their overall thermal output. However, in most reactors the rate of conversion of ^{238}U to ^{239}Pu lags behind the rate at which the fissile isotopes are split, requiring eventual removal of the spent fuel while it still contains abundant ^{238}U. The net result of this once-through strategy is that most of the ^{238}U is indeed wasted.

It is possible to make much more efficient use of ^{238}U, however, by using a different approach to nuclear reactor design. Most current reactors employ "thermal" neutrons, which have been slowed by the moderating influence of water. Slowing the neutrons greatly increases the probability of their being captured by ^{235}U, leading to fission and the production of additional neutrons. The main advantage of this approach is that the fuel needs to be enriched to only about 4 percent ^{235}U. Fission can also be sustained using unmoderated, more energetic fast neutrons, but this approach requires greater enrichment of fissile material, to about 20 percent of the fuel, which substantially increases costs.

However, the absorption of fast neutrons by ^{238}U also leads to more rapid production of ^{239}Pu, with the net result that a fast neutron reactor can be designed to produce (or "breed") a greater amount of fissile fuel than it consumes. The spent fuel from such a reactor can then be reprocessed to separate the newly produced ^{239}Pu, which is recycled as fissile fuel. This mode of operation can in principle make productive use of most of the originally mined ^{238}U, dramatically extending the potential magnitude of useful uranium resources and leaving far less waste to be stored. The remaining wastes would mostly consist of fission products that decay in a matter of centuries (Figure 14.9), eliminating the need to build repositories that are more enduring than the pharaohs' tombs.

Unfortunately breeder reactors also have a dark side: Concentrated ^{239}Pu can be used in the construction of nuclear weapons. Routine reprocessing of fuel for nuclear power generation might therefore provide opportunities for the proliferation of nuclear weapons capability. The combination of higher cost and greater security risk has led the United States away from fast reactors and reprocessing as a matter of national policy. In contrast, France and Kazakhstan have

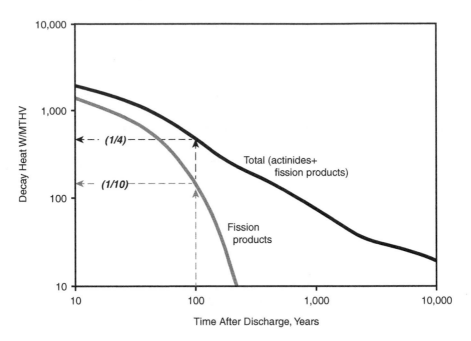

Figure 14.9. Decay heat in spent nuclear fuel that has been irradiated to 50 GWd/MTIHM.

Note that both horizontal and vertical scales are logarithmic; each tick mark denotes a factor of 10 difference. The lower gray curve represents the sum of fission products, and the upper black curve represents total heat. Note that heat from fission products has decreased by approximately 90 percent after one century and is negligible after two centuries. Total heat declines approximately 75 percent after one century, but then declines more slowly for thousands of years afterward (modified from Wigeland et al., 2006).

both operated commercial fast neutron reactors in the past, and several other countries are developing them. Ironically one of the chief impediments to their development has been relatively low uranium prices, which have kept nuclear fuel efficiency a low priority.

Conclusions

Deep geologic burial of energy wastes holds considerable appeal due to the potential to sequester harmful materials away from the Earth surface environment for long periods. Returning these materials to their point of origin also provides a certain sense of symmetry and effectively removes them from our consciousness. However, putting energy wastes back into the ground may not always be as easy as taking the original energy resources out.

At present, the only energy-related wastes to be routinely returned underground are saline waters produced with oil and gas. Deep reinjection generally has induced small earthquakes in some areas but generally seems to cause few problems.

Decades of subsurface injection of CO_2 for the purpose of enhanced oil recovery have amply demonstrated its feasibility. Collecting and injecting enough CO_2 to make a noticeable impact on the Earth's atmosphere will pose a bigger problem. Current projects use CO_2 from high-concentrations point sources such as coal-fired power plants, but these collectively account for only about half of current emissions. Disposal of larger emissions fractions would therefore require massive changes in our current emission patterns, for example, the conversion of transportation from direct fossil fuel combustion to electricity. Disposing of CO_2 will also be expensive. These costs may be justifiable if they help to avert catastrophic atmospheric change, but at the moment there is little consensus on who would pay for carbon capture and storage.

Deep geologic disposal of nuclear wastes seems far less problematic by comparison, at least from geological and cost perspectives. In part this reflects the immense concentration of energy within atomic nuclei; a large amount of power can be produced from a relatively small amount of fuel. It appears that a variety of different geological configurations can safely contain radioactive wastes, and that the natural mobility of such wastes is inherently limited. The major technical issues appear to be the potential for local groundwater contamination and the requirement that geologic repositories remain intact and undisturbed for almost incomprehensibly long periods.

For More Information

Archer, D., 2005, Fate of fossil fuel CO_2 in geologic time: *Journal of Geophysical Research*, v. 110, C09S05, doi:10.1029/2004JC002625.

Bentridi, S.-E., Gall, B., Gauthier-Lafaye, F., Seghour, A., and Medjadi, D.-E., 2011, Inception and evolution of Oklo natural nuclear reactors: *Comptes Rendus Geoscience*, v. 343, p. 738–748.

Chadwick, R. A., and Noy, D. J., 2010, History-matching flow simulations and time-lapse seismic data from the Sleipner CO_2 plume: *Petroleum Geology Conference Proceedings*, v. 7, p. 1171–1182.

Cochran, T. B., Feiveson, H. A., Patterson, W., Pshakin, M. V., Ramana, M. S., Suzuki, T., and von Hippel, F., 2010, Fast Breeder Reactor Programs: History and Status: Princeton, NJ, International Panel on Fissile Materials and Princeton University, Research Report 8, 115 p.

Dooley, J. J., Dahowski, R. T., Davidson, C. L., Wise, M. A., Gupta, N., Kim, S. H., and Malone, E. L., 2006, Carbon Dioxide Capture and Geologic Storage – A Core Element of a Global Strategy to Address Climate Change: College Park, MD, Battelle, Joint Global Change Research Institute, 37 p.

Frohlich, C., Potter, E, Hayward, C., and Stump, B., 2010, Dallas–Fort Worth earthquakes coincident with activity associated with natural-gas production: *The Leading Edge*, March, p. 270–275.

Hanks, T. C., Winograd, J., Anderson, E. R., Reilly, T. E., and Weeks, E. P., 1999, Yucca Mountain as a radioactive waste repository: *U.S. Geological Survey Circular* 1184, 19 p.

Healy, J. H., Rubey, W. W., Griggs, D. T., and Raleigh, C. B., 1968, The Denver earthquakes: *Science*, v. 161, p. 1301–1310.

Hess, H. H., Adkins, J. N., Benson, W. E., Frye, J. C., Heroy, W. B., Hubbert, M. K., Russell, R. J., and Theis, C. V., 1957, *The Disposal of Radioactive Waste on Land*: Washington, DC, National Academy of Science-National Research Council, 142 p.

Hubbert, M. K., and Rubey, W. W., 1959, Role of fluid pressure in mechanics of overthrust faulting: I. Mechanics of fluid-filled porous solids and its application to overthrust faulting: *Geological Society of America Bulletin*, v. 70, p. 115–166.

International Atomic Energy Agency, 2003, *The Long Term Storage of Radioactive Waste: Safety and Sustainability – a Position Paper of International Experts*: Vienna, International Atomic Energy Agency, 18 p.

International Energy Agency, 2012, *CO2 Emissions from Fossil Fuel Combustion – Highlights*: Paris, Organization for International Cooperation and Development/International Energy Agency, 125 p.

IPCC, 2005, *IPCC Special Report on Carbon Dioxide Capture and Storage. Prepared by Working Group III of the Intergovernmental Panel on Climate Change* [Metz, B., O. Davidson, H. C. de Coninck, M. Loos, and L. A. Meyer (eds.)]: Cambridge, UK, and New York, Cambridge University Press, 442 pp.

Kelemen, P. B., Matter, Juerg, Streit, E. E., Rudge, J. F., Curry, W. B., and Blusztajn, J., 2011, Rates and mechanisms of mineral carbonation in peridotite: Natural processes and recipes for enhanced, in situ CO_2 capture and storage: *Annual Reviews of Earth and Planetary Sciences*, v. 39, p. 545–576.

Kraft, T., Mai, P. M., Wiemer, S., Deichmann, N., Ripperger, J., Kästli, P., Bachmann, C., Fäh, D., Wössner, J., and Giardini, D., 2009, Enhanced geothermal systems: Mitigating risk in urban areas: *Eos*, v. 90., p. 273–280.

Massuchusetts Institute of Technology, 2003, *The Future of Nuclear Power – an Interdisciplinary MIT Study*: Cambridge, Massachusetts Institute of Technology, 170 p.

Meshik, A. P., 2005, The workings of an ancient nuclear reactor: *Scientific American*, November, p. 82–91.

National Energy Technology Laboratory, 2010, *Carbon Dioxide Enhanced Oil Recovery – Untapped Domestic Energy Supply and Long Term Carbon Storage Solution*: Washington, D.C., U.S. Department of Energy, 32 p.

National Energy Technology Laboratory, 2012, *Carbon Utilization and Storage Atlas*, 4th Ed.: Washington, DC, U.S. Department of Energy, Office of Fossil Energy, 129 p.

National Research Council, 2013, *Induced Seismicity Potential in Energy Technologies*: Washington, DC, National Academies Press, 248 p.

Solomon, S., 2007, *Carbon Dioxide Storage: Geological Security and Environmental Issues – Case Study on the Sleipner Gas Field in Norway*: Oslo, Bellona Foundation, 126 p.

U.S. Department of Energy, 2002, Environmental Impact Statement for a Geological Repository for the Disposal of Spent Nuclear Fuel and High-Level Radioactive Waste at Yucca Mountain, Nye County, Nevada – Reader's Guide and Summary: Washington, DC, U.S. Department of Energy Office of Civilian Radioactive Waste Management, 103 p.

Waltar, A. E., and Reynolds, A. B., 1981, *Fast Breeder Reactors*: New York, Pergamon Press, 853 p.

Wigeland, R. A., Bauer, T. H., Fanning, T. H., and Morris, E. E., 2006, Separations and transmutation criteria to improve utilization of a geologic repository: *Nuclear Technology*, v. 154, p. 95–106.

Zoback, M. D., 2012, Managing the seismic risk posed by wastewater disposal: *Earth Magazine*, April, p. 38–42.

15

How Long Is Forever? Energy and Time

> Prediction is very difficult, especially about the future.
>
> *Commonly attributed to Niels Bohr*

Picture a herd of woolly mammoths grazing amid grass-covered, rolling hills on an autumn morning. They are moving slowly along a river that is muddy with silt, derived from melting of the last remnants of a glacial ice sheet that once covered half the continent. There is a chill in the air, with a strong breeze blowing in from the northwest. A group of fur-clad hunters stalks the mammoths from the opposite direction, discussing details of their attack plan with hand signals and low guttural grunts. If the hunt is successful they may survive through the long winter ahead; otherwise their future is in doubt.

Although incredibly primitive by modern standards, this scene could actually have played out as recently as ten thousand years ago, hardly a blink of the eye in geological terms. Agriculture began to spread through the eastern Mediterranean region only about two thousand years afterward, and the earliest pyramids in Egypt were built at the approximate midpoint in time between the disappearance of woolly mammoths and the appearance of cell phones.

The historical details have been lost to the ages, but it seems clear that massive environmental change, brought on by a respite in continental glaciation, set the stage for early human civilization. If Stone Age and earlier humans had any thoughts at all about the future of energy, they would almost certainly have been along the lines of "How can I keep warmer?" The idea that global warming might eventually become a problem would have been utterly unimaginable.

Since that time, the cutting edge of technology has advanced from stone axes to lasers, and our dominant energy sources have evolved from human muscle to fossil fuels. The historically recent emergence of the latter has transformed the world at unprecedented rates, leading many to worry about how long we can maintain the standard of living that we have come to accept as our birthright. Equally important, how we can extend this standard to the large proportion of the world's population who presently are being left behind? And what will be the eventual environmental price of our energy use? Underlying all these questions is a more fundamental one that is often left unspoken: How much of the future do we realistically care about?

The Foreseeable Future

Most humans are strongly invested in the events of the next weekend, but probably do not care at all about the next major reshuffling of the Earth's tectonic plates or the eventual extinction of the Sun (an exception is made for professors and students of geology and astronomy). Between these two extremes, our time frame of interest is strongly skewed by the duration of our own expected lifetimes, plus perhaps the lifetimes of our children or grandchildren. It is considerably harder to think in detail about the future of our great-grandchildren, whom we may never even meet, and the generations beyond that really are just strangers.

Our financial concerns tend to be even shorter-lived. The most ambitious personal financial plans aim to pay off the mortgage and invest for retirement, goals that typically require thirty to forty years of foresight at most. Large corporate or public projects typically don't look much further ahead than this, because money available in the future is worth less than money available right now. The future is literally "discounted" by its uncertainty. This is why lottery winners receive much less money if they elect an immediate lump-sum payment, rather than a series of yearly payments made in the future. At some point the future effectively ceases to have any economic value at all, greatly limiting our desire to invest in it. For example, at a 7 percent annual discount rate, $1 million paid 100 years from now is presently valued at only about $1,200, a total discount of 99.99 percent.

Taken together these considerations suggest that humans as a race are mostly only concerned with what happens over the next hundred years or so, a very brief period relative to the history of human civilization. If this is true, our expectations of the future are surprisingly modest. The goal is simply to hang on for one more century. After that who knows what will happen? Perhaps the Earth will be devastated by an asteroid impact, and the few remaining survivors will live in colonies on Mars? Or maybe the depletion of the Earth's resources will force a decline into either a dystopian society or simple anarchy, as imagined by numerous science fiction authors and Hollywood screenwriters? The probability of any of these scenarios may be low, but they serve to illustrate the difficulty of predicting the distant future.

Countering this rather cynical and selfish perspective, human population will probably still be very large in the twenty-second century, and most if not all of these people will probably still want to live on Earth. We may therefore feel a certain moral responsibility to steward the Earth's resources for the longer-term good of our species. The details of how this stewardship should be done are a bit fuzzy, though. Ideally we should strive to balance energy use with natural rates of energy renewal or replenishment. Energy renewability by itself is not enough, however; we must also ensure access to sufficient *power*, defined as the rate at which we use energy. Energy supply is an important but somewhat esoteric quantity, of little concern to everyday life. Power is what keeps the lights on.

Fossil fuels are all about power; they have the proven ability to provide energy at virtually any rate we desire. Their ability to continue doing so over the next century (or longer) may be questioned, however, since their rate of usage vastly exceeds their natural rate of renewal. The consequences of their use are no longer in question, at least not among the scientific community. The Intergovernmental Panel on Climate Change (IPCC) has issued a series of increasingly urgent and well-documented reports, expressing the scientific consensus that combustion of fossil fuels has caused warming of the atmosphere. The remaining questions revolve around the future magnitude of greenhouse gas emissions, and how the Earth's complex natural systems will respond to these emissions.

Regardless of the urgency of atmospheric warming, economic and political realities will make it very difficult to reduce fossil fuel use in the near term, let alone eliminate it. The relatively low monetary cost of obtaining power from fossil fuels lies at the core of these realities. Their worst environmental costs appear to lie in the future and therefore tend to be discounted. The magnitude of remaining fossil fuel supplies will therefore directly influence both our future access to cheap power and the severity of eventual environmental change. But what is this magnitude? This question is highly contentious and will be the subject of most of the remainder of this chapter. Before considering the future of fossil fuels, it is first necessary to develop a basic understanding of their past usage.

The Evolution of Fossil Fuels

Probably no one alive at the beginning of the 19th century could have predicted the dramatic changes that would be brought about by cheap energy and new technology. The two have been continually intertwined, each enabling changes in the other. Our current understanding of this relationship did not spring into being overnight, however. It has evolved through several distinct generations, or eras, of energy use, which might be likened to the major eras of geologic time.

As described in Chapter 2, the past 541 million years of geologic time is subdivided into three eras, based on the types of fossils found within the rocks of each. The boundaries between geologic eras coincide with especially profound changes, as indicated by the sudden disappearance of large numbers of preexisting fossils, followed by the slow appearance of new fossil types not seen before. The oldest era is the Paleozoic, derived from Greek roots meaning "old" and "animal." The next is Mesozoic, for "middle," and the last is Cenozoic, for "new." The first two eras ended in mass extinctions, resulting in the loss of more than 90 percent of marine organisms at the end of the Paleozoic and of the dinosaurs at the end of the Mesozoic. Many groups also survived these extinction episodes, however. For example, sharks, ferns, and conifers all emerged during the Paleozoic and continue to enjoy great success today.

A similar scheme can be applied to our much shorter history of fossil fuel use, which has also advanced in discrete stages. Since the defining characteristic of fossil fuels is their ability to produce heat, we might appropriately substitute the suffix -*thermic* in place of -*zoic*. In parallel with the geologic timescale, this newly proposed scheme includes three major eras: the Paleothermic, the Mesothermic, and the Cenothermic.

During the Paleothermic era of fossil fuel use, coal was king. Its close cousin oil shale emerged at about the same time but has yet to make much of an impact. Coal and oil shale were the first fossil fuels to be used because they occur in abundance at or near the Earth's surface and thus are easy to find and extract. Both have been known for many centuries. The late 18th to early 19th century development of external combustion steam engines provided a means of transforming heat into motion and thereby enabled a boom in the use of coal.

318   Geofuels: Energy and the Earth

The Mesothermic era can be conveniently subtitled "The 20th Century." It began with the widespread adoption of liquid fuels refined from crude oil, and the parallel development of internal combustion engines. The key to the success of the Mesothermic era is natural concentration of oil and gas into rich, localized accumulations. Because they lie beneath the Earth's surface, Mesothermic oil and gas fields are harder to find than coal and oil shale, but this difficulty is far outweighed by the enhanced utility of naturally fluid fuels. Once found, they are relatively easy to exploit. The rise of crude oil and natural gas did not drive coal to extinction, but it did demote it to applications where low cost is paramount, and solid fuel does not pose a major impediment.

The Cenothermic era arrived largely unheralded in the late 20th century but has since revolutionized our perceptions of fossil fuels. The characteristic feature of Cenothermic oil and gas is their occurrence in disseminated deposits spread across wide areas. This includes oil and gas that have never strayed far from their source rocks, and are locked in place by permeability limitations (the subject of Chapter 10), It also includes oil and gas that did move away from its source, but that has since been immobilized (the subject of Chapter 11). In both cases extraction is more difficult than for Mesothermic deposits, requiring many more wells spread out across relatively large areas. Even then, additional help is needed for these deposits to flow to a well bore, such as horizontal drilling, hydraulic fracturing, or steam injection.

Peak Oil: The End of an Era?

As important as it has been, the Mesothermic era appears to be drawing to a close. Its reliance on discrete natural concentrations of oil and natural gas implies that once found, those concentrations can be drained relatively quickly and easily. This basic fact enabled M. King Hubbert to predict in 1956 that the United States, which had already discovered most of its largest fields, would soon see its oil production reach a peak and then begin to decline. The central tenet of his "peak oil" concept was that the decline of U.S. production would be a mirror image of its rise, eventually resulting in a nicely symmetrical bell-shaped curve (Figure 15.1). If you know the shape of the rising limb of such a curve, along with the total amount of available reserves, then the timing of peak production can be easily deduced. Hubbert predicted that a peak of U.S. production would arrive in either 1965 or

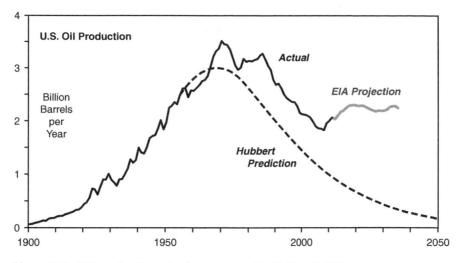

Figure 15.1. U.S. crude oil production compared to Hubbert's 1956 prediction (data from the U.S. Energy Information Administration, http://www.eia.gov).

1970, on the basis of two different reserves estimates. The latter prediction ultimately proved correct.

Global *discoveries* of new oil and gas accumulations peaked in the 1960s to 1970s, leading many experienced and thoughtful experts to predict that a peak in world oil production would soon follow. To date, reality has stubbornly refused to cooperate with these predictions. Unlike oil production in the United States, world oil production has not peaked (Figure 15.2). Production in 2012 exceeded that of any previous year, and the International Energy Agency projected that increases would continue until at least 2040. Even Hubbert's apparent success in predicting U.S. peak oil production may not have been what it seemed; after decades of steady decline U.S. oil production actually reversed course and began to rise again in about 2008.

So why has world oil production thus far failed to peak as expected? Possibly the peak has only been delayed and is just around the corner. Only time will tell. Meanwhile the peak oil concept itself needs to be more closely scrutinized. As it turns out, real-world oil production histories often fail to conform to the idealized bell-shaped curve, a fact first noted by Hubbert himself. Conversely, bell-shaped production histories do not necessarily reflect resource depletion. For example, anthracite coal production in Pennsylvania traced out a beautifully symmetric curve during the late 19th and early 20th

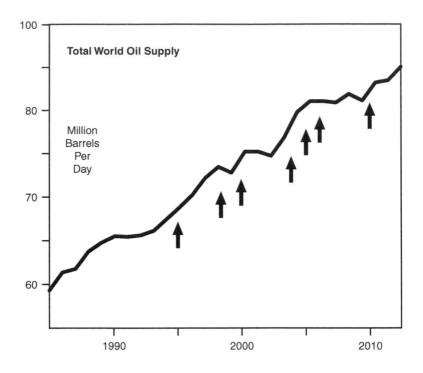

Figure 15.2. Recent world oil production history. Arrows indicate the timing at which various authors have predicted that world oil production would peak (data from U.S. Energy Information Agency, http://www.eia.gov).

centuries, peaking around 1920. As geologist Peter McCabe pointed out in 1998, however, nearly all of the known anthracite remains untouched! Rather than reflecting resource depletion, the decline in production resulted from the entry of cheaper energy sources into the market, including oil and lower-rank coal.

Although intuitively appealing, the peak oil concept entirely disregards the fact that the production of any natural resource is controlled as much by market factors and extraction technology as it is by natural abundance. Whereas the Earth's total oil endowment is clearly fixed, its economically extractable endowment of oil is not.

Held in Reserve

The word "reserves" is frequently used to indicate a quantity of crude oil or other resource remaining in the ground, usually with the implication that this quantity represents a finite limit. But what does the word really mean? Its actual definition is often left unspoken, and the common addition of modifying terms such as "proven" or "probable" only multiplies this confusion.

The word "reserves" derives from the Latin verb *reservare*, which means to keep back. If you reserve a hotel room your expectation is that a room will be held for your arrival; in military usage the noun "reserves" refers to troops who are held back from the main fighting force. The word "reservoir" shares a similar derivation; for example, an artificial lake can be thought of as part of a river that is being kept back by a dam. Note that in none of these cases do reserves imply an absolute quantitative limit. The hotel has other rooms in addition to the one you reserved, the military has many active duty soldiers in addition to reserves, and the whole volume of water that flows through a river is not contained in a single reservoir.

Extending this idea to crude oil, reserves would constitute oil that has been intentionally held back, as a buffer against future increases in demand or unexpected supply interruptions. If so, then there is really not much difference between oil and any other commodity. For example, Peter McCabe used an analogy to baked beans kept in a grocery store warehouse. The purpose of the warehouse is to ensure that grocery store shelves can be kept continuously stocked, regardless of short-term fluctuations in the public's appetite. Should the shelves become depleted, the store manager has only to call the warehouse and ask them to send over some additional cases.

The amount of baked beans stored in the grocery warehouse obviously has no bearing on the total amount of baked beans that exist in the world at a given moment or that may exist in the future. The warehouse is merely a tool used by grocery stores to guarantee a steady supply of beans to store shelves. With proper management the warehouse need never actually run out of beans; the cans that are removed to stock store shelves are continuously replenished with shipments from the baked bean supplier. Enough beans need to be kept on hand to meet normal fluctuations in demand, but no more. Increasing these reserves would only add cost, because of the need for a bigger warehouse.

A key requirement of an effective warehouse is that its contents be readily accessible at a moment's notice; otherwise it cannot serve its primary function. One might therefore expect the crude oil equivalent of the grocery store warehouse to be storage tanks and pipelines, from which oil can be tapped at a moment's notice. To a certain extent this is true; storage tanks do provide some buffer against day-to-day fluctuations in supply and demand. The total volume of oil contained in tanks and pipelines is quite small, however; for example, according to Energy Information Administration data U.S. oil stocks total only about 200 million barrels (http://www.eia.gov). That is enough to satisfy U.S. demand for about ten days at most, assuming every drop could be used. Tanks and pipelines provide no protection against supply fluctuations that occur over longer periods.

The United States does store larger quantities of oil in its Strategic Petroleum Reserve, which consists of a series of depleted fields in the Gulf Coast region. The federal government purchases crude oil to pump back into these fields, with the idea that it can be quickly pumped out again when needed. This reserve was established in the aftermath of the OPEC oil embargo of 1973–1974, to ameliorate the effects of future supply disruptions or price shocks. When filled, it holds enough oil to meet the total needs of the United States for about one month, or to replace all imported oil for about two months. Recently its use has largely been limited to dealing with regional supply disruptions caused by hurricanes. Many other countries and companies have established their own strategic reserves, which according to the U.S. Energy Information Administration total about 4 billion barrels worldwide. This equals about 1.5 months of world oil consumption at current rates.

Reserves on Demand

The vast majority of our available crude oil does not reside in tanks, pipelines, or strategic petroleum reserves. Instead it remains in its original state, trapped below the surface within porous rocks. Even so, these geologic reserves can still be viewed as analogous to baked beans in a grocery warehouse, provided you take a slightly longer-term view. Increases in the market demand for oil can be accommodated by increasing the rate at which it is pumped from the ground. Accomplishing such increases generally requires more than just opening a valve, however.

Returning for a moment to the grocery analogy, imagine a warehouse that was originally designed with a very small delivery chute, through which beans must be pushed out one can at a time. The chute limits the rate at which store shelves can be restocked. It may be large enough to keep stores supplied on most days, but should holiday picnics trigger unusually high bean demand the small chute could cause a temporary shortage. A carpenter could remedy the situation by building a larger chute, but that job could take days or maybe even weeks. By then the holiday would be only a distant memory.

The narrow well bores through which crude oil must pass are the functional equivalent of small warehouse chutes. Minor increases in oil demand may be accommodated by pumping it more quickly from existing wells, but larger increases require drilling more wells. Drilling those wells can easily require months or years to complete, provided that sufficient drill rigs are available and work starts immediately. Meanwhile the failure of oil production to keep pace with consumer demand may result in rising oil prices, especially if the increased demand reflects a global trend. Eventually, though, the higher prices should provide an incentive for greater oil production, which in principle should push prices back down.

Proven reserves of crude oil definitely do *not* equal the total amount of potentially accessible oil contained in the Earth's crust. The history of reported reserves in the United States makes this point clear. For example, in 1921 the United States had approximately 8 billion barrels in proven reserves. By midcentury this figure had grown by a factor of nearly 20. In 1956 Hubbert estimated U.S. "ultimate potential oil reserves," the sum cumulative production, proven reserves, and future discoveries, to fall between 150 and 200 billion barrels. By 2010 cumulative oil production had eclipsed the latter amount, and production rates were on the rise.

A closer examination reveals that not only has the known original endowment of recoverable oil in the United States increased through time, it has done so in a remarkably reliable fashion. The upper curve in Figure 15.3 depicts cumulative additions to proven reserves, which include both the discovery of new oil fields and improved recovery from existing fields. This total rose inexorably throughout the 20th century, with steeper segments corresponding to large discoveries or other upward revisions. The most notable of these was the 1969 discovery of the Prudhoe Bay field in Alaska, the largest single oil field ever discovered in the United States.

The most striking aspect of U.S. petroleum history is how closely additions to reserves have paralleled production. The vertical difference between the two curves in Figure 15.3 represents the actual oil reserves that existed during any given year, expressed in barrels of oil. This number increased from approximately 3 billion barrels in the early years of the 20th century to a peak of 35 billion barrels in 1970,

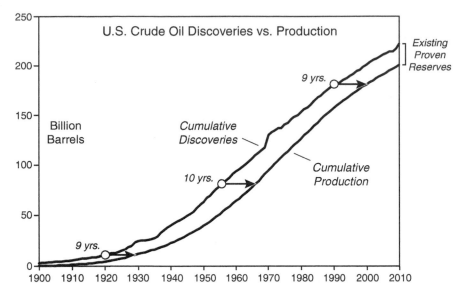

Figure 15.3. Cumulative crude oil reserves and production histories for the United States.Horizontal arrows indicate the number of years of actual future production that were available as reserves at the times indicated by the circles (modified from McCabe, 1998; data from U.S. Energy Information Administration, http://www.eia.gov).

and then dropped to about 17 billion barrels in 2008. As of this writing it has risen again. It is more illuminating, however, to compare the two curves in terms of their offset in years, because this measures how long reserves can be expected to last if no new additions are made. For example, the reserves in hand in 1920 were sufficient to meet actual domestic oil production only through 1929. After the Prudhoe Bay discovery proven reserves were sufficient to satisfy actual production through 1981.

Surprisingly, ever since the early 20th century the United States has *always* had proven oil reserves equal to between eight and eleven years of actual future production! New reserves have continuously been added at a rate that almost exactly matches the rate at which old reserves have been depleted, so that we have never "run out." Although the economic mechanisms supporting this balance are undoubtedly complex, the net result has been almost shockingly consistent. Furthermore, the same balance was maintained both as total domestic production increased prior to 1970 and as it decreased afterward.

In contrast to the conventional view that production follows from the discovery of new reserves, the reverse actually appears to be true. Reserves are developed in response to the market demand for oil and function in a manner remarkably similar to the inventory kept in a grocery warehouse.

Oil Reserves: The Larger View

U.S. oil production did decline for several decades following its 1970 peak, but apparently not because reserves were running out. Instead, this decline might be compared to a grocery store chain that has lost market share to a competing discount brand. The discount brand in this case was cheaper imported oil, which began to replace domestic production. As domestic production shrank so did domestic reserves, for the same basic reason that a failing grocery store chain might downsize its warehouse. Decreased domestic production did not lead to shortages, though; in fact, the opposite happened. The availability of cheap foreign oil propelled domestic consumption to new heights.

The United States contains more than half of the world's producing oil and gas wells, and this greater density of drilling has long made it the bellwether for development patterns elsewhere. As in the United States, estimated cumulative additions to world reserves grew progressively through the latter half of the 20th century and into the early 21st (Figure 15.4). Unlike in the United States, however, cumulative additions to world reserves appear to have actually grown *faster* than cumulative world production. Reserves in 1951 were sufficient to satisfy actual production through 1964, a situation comparable to that in the United States. However, 1982 estimated proven reserves equaled 28 years of actual future production, and it appears that this time interval has continued to increase since.

The keyword with respect to proven world reserves is "estimated." Reserves figures reported by publicly traded companies in the United States are closely scrutinized by the Securities and Exchange Commission to ensure that shareholders are not being defrauded. No such assurances exist for the reserves reported by nationalized oil companies outside the United States. These companies, which are believed to hold more than 90 percent of the world's oil reserves, are free to announce any figures they wish. Without access to the underlying drilling and production data it is impossible to know for sure whether the reported reserves are accurate. Inflated estimates have been suspected

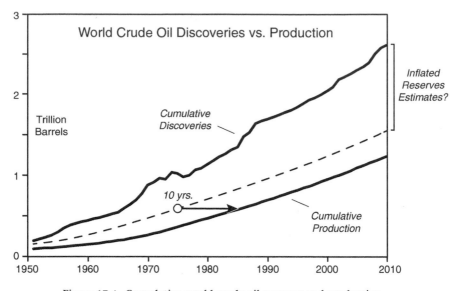

Figure 15.4. Cumulative world crude oil reserves and production histories.Dashed curve indicates a hypothetical case in which actual reserves at any given time are sufficient to satisfy ten years of future production (data from BP Statistical Review of World Energy, http://www.bp.com/en/global/corporate/about-bp/energy-economics/statistical-review-of-world-energy.html, and from the Oil and Gas Journal).

in a number of cases, for example, the sudden increases in Saudi reserves that were reported in the late 1980s. More recently, large reserves additions announced by Iran and Venezuela have also drawn critical attention. In all of these cases reported reserves appear to have been increased without a corresponding increase in drilling.

Even if non-U.S. reserves have been grossly exaggerated, this fact by itself does not necessarily imply an impending world shortage of crude oil. So long as the reserves "warehouse" is restocked at a rate fast enough to keep pace with oil production, the actual size of proven reserves at any given time is irrelevant. The dashed line in Figure 15.4 illustrates a hypothetical case in which cumulative global reserves have been continuously maintained at a size adequate to meet about ten years of future production, in analogy to the observed history in the United States. Reported world reserves for 2010 exceed this hypothetical case by a factor of 4. This may reflect inflated estimates, but so long as *true* proven reserves do not fall below the dashed line there is no possibility of real shortage.

The Resource Pyramid

How is it possible for reserves of crude oil to grow continuously when we know that the total amount of oil contained in the Earth is finite? The short answer is that reserves only constitute a small fraction of the Earth's natural endowment of crude oil. This relationship can be conveniently expressed through the "resource pyramid," a commonly used metaphor for crude oil and other geologic resources (Figure 15.5). The overall volume of the pyramid represents the Earth's total endowment of oil, whereas its uppermost tip represents oil reserves. Geologists define reserves as *the known amount of crude oil that can be profitably extracted, assuming present-day technology and economic conditions.* Note that under this definition, the price of oil matters as much as its physical abundance in the Earth. If the price of oil rises, it becomes profitable to dig deeper into the pyramid, converting more of the Earth's total oil endowment into reserves. The same thing can happen if a new technology, for example, rotary drilling or hydraulic fracturing, allows oil to be extracted at a lower cost.

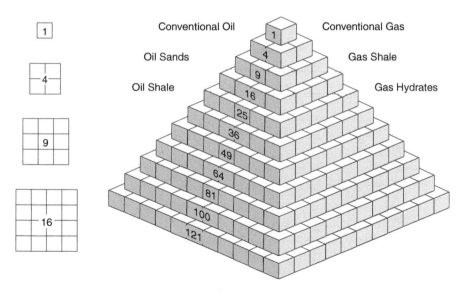

Figure 15.5. Schematic illustration of the resource pyramid for oil and gas. Note that the volume of each layer (= number of blocks) increases geometrically moving downward in the pyramid, as resource quality decreases. The true shape and ultimate volume of the pyramid are unknown, however (modified from McCabe, 1998, and others).

In addition to being profitable to extract, reserves by defini-tion must be *known*. The subdivisions proven, probable, and possible denote the certainty with which they are known to exist. For oil and gas this knowledge is gained primarily from drilling; more wells pro-vide more certainty. Strange as it may seem, the vast majority of the Earth's crust has never been tested by drilling, and much of it has never even been imaged by seismic surveys. The reason for this knowl-edge gap is simple: drilling and seismic surveys are extremely expen-sive. Oil producers must therefore weigh new exploration against the anticipated future price of oil, and the competing value of other pos-sible investments. Only the most promising areas have been drilled.

Our comprehension of oil reserves also changes continuously as geologists learn more about how these deposits are formed. In the mid-19th century such knowledge was extremely limited, and oil was found mostly by digging in places where it naturally seeps out onto the Earth's surface. The anticlinal theory of oil entrapment came along later in that century. It eventually resulted in an enor-mous increase in known commercial oil supplies, including most of the oil in the Middle East. Later still it was discovered that oil could be trapped between different rock layers that gently converge in the subsurface, even if those layers do not form an anticline. These more subtle "stratigraphic" traps further expanded known supplies. Over the past few decades we discovered that oil and gas can also be prof-itably extracted from rocks that lack any distinct trapping mechanism. These Cenothermic deposits are held in place by the low permeability of their host rocks, or by thickening of oil due to microbial scavenging of its solvent components.

Reserves figures also depend directly on the technology used to find, extract, and utilize crude oil. In 1800 the world's total reserves were near zero, because oil was not yet widely used as an energy source and extracting it from the ground was very difficult. Shovels and pickaxes were used to dig the earliest oil wells, eventually giv-ing way to crude drilling tools lowered by cables, and later to mod-ern rotary drilling. Each new advance has spurred increases in world oil reserves, which up until now have conveniently kept pace with world's increasing population and their demand for oil.

Can this pattern be sustained forever? Clearly not, unless "for-ever" stretches into the millions of years required for fossil fuel renewal. But how much oil is available over timescales that humans care about? The answer to this question depends on how much further down we can profitably dig into the resource pyramid.

Digging Deeper

The next level down in the pyramid contains "resources," which include that portion of Earth's total endowment of coal, oil, or natural gas that we think could potentially be converted into reserves at some time in the future. In 1972 the director of the U.S. Geological Survey, V. E. McKelvey, proposed that resources are of three basic types: those that have not yet been discovered, those that cannot be profitably extracted under present-day technology and economic conditions, and those that haven't been discovered *and* are not presently profitable. The future conversion of resources into reserves depends on new exploration efforts, advances in technology, increased commodity price, or some combination of these factors.

Because of this fundamental dependency on events that have not yet happened, the ultimate magnitude of fossil fuel resources is effectively impossible to know. You might as well ask what home mortgage rates will be in ten years, or which stocks will rise the most in value. The futility of the latter question was nicely demonstrated by a recent experiment in which a British house cat named Orlando picked stocks by throwing a toy mouse at a list of companies. Over the course of a year Orlando's picks outperformed those made by a team of market professionals by a large margin. Either Orlando possessed a level of innate financial acumen that eludes most humans, or else future stock prices are governed by information we do not yet possess. Foreknowledge of the impact of future fossil fuel technology and markets has proven to be similarly elusive.

Neither reserves nor resources represent fixed, finite quantities, but reserves are at least objectively measurable, whereas resources are by definition subjective. As described by the MIT economists M. A. Adelman and M. C. Lynch in 1997, reserves represent "inventory" that has been created and renewed by continuous investment in exploration and development. Resource estimates, on the other hand, represent "implicit forecasts of investment, therefore of future technology, which nobody knows." In the case of crude oil and natural gas these forecasts have generally grown progressively more optimistic through time, resulting in ever-larger resource estimates.

The mutable nature of reserves and resources is perhaps best illustrated by the fact that they can also decrease. For example, if oil prices drop precipitously, oil that previously could have been recovered at a profit might instead be left in the ground. This actually happened during the late 1980s through 1990s, when low prices led to the abandonment of many marginally productive wells. Coal provides an even more dramatic example. In 1956 Hubbert estimated that the United States held 950 billion metric tons of recoverable coal, but more recent estimates have shrunk to about half that amount. Production of coal during the intervening period amounted to about 41 billion tons, far too little to account for the discrepancy.

The missing 400 billion tons of U.S. coal did not simply vanish; it remains right where it has always been. What has changed is our perception of economic recoverability. During the first half of the 20th century most coal was mined underground, a relatively expensive and labor-intensive endeavor. Surface mining accounted for only about a third of U.S. coal production in the 1950s, but by the late 1970s the surface proportion had increased to more than two-thirds. Advances in mechanization made it much cheaper to mine coal from the surface than from underground mines; that development in turn made it harder for higher-priced underground coal to compete. Over time, this change recalibrated our perception of the economics of coal mining, to the detriment of some formerly profitable underground coal. Ironically, the availability of more cheaply mined surface coal has contributed to downward revisions in overall resource estimates.

The Energy Cost of Energy

One of the oldest truisms in business is that "you must spend money to make money." This truism certainly applies to the energy business, where multibillion dollar investments are routinely made in drilling, refining, mine development, power plant construction and operation, and other processes. Part of the cost of these projects can be traced to the energy required to carry them out, which means you must also "spend energy to make energy." For example, drilling an oil well requires energy to turn the drill bit, to pump the drilling mud that cools the bit and removes rock chips, and to raise the drill pipe out of the hole. More broadly, energy is needed for hundreds of other steps required to complete a well, ranging from the mining of iron ore for steel to propelling the vehicles that carry drillers to the well site.

Collectively this adds up to a lot of energy that must be expended in order to find and produce crude oil, but the energy contained in the oil itself easily pays back the initial energy expenditure. Geographer Cutler Cleveland, together with coauthors Robert Costanza, Charles

Hall, and Robert Kaufmann, argued in a 1984 paper in the journal *Science* that this favorable energy payback ratio was a major driving force on the United States and other national economies. They called it "energy return on investment," EROI for short, which is defined as "the ratio of gross fuel extracted to the economic energy required to deliver the fuel to society in a useful form." The EROI for conventional oil fields discovered during the early 20th century has been estimated to be as high as 100, meaning that for every 1 unit of energy invested, 100 units were returned. Cleveland estimated that the ratio had fallen to something in the range of 10–20 by the end of the century, still very good compared to that of other energy sources.

EROI is important because it directly impacts the rate at which we will use up our total energy supplies. Energy consumption can be thought of in two parts: the useful energy that can satisfy practical needs such as transportation and lighting and the additional energy that must be invested to *furnish* that useful energy. For example, let's say that we need 1 unit of useful energy and must invest another 0.11 unit to get it, which totals a gross energy extraction of 1.11 units. The EROI in this case is 1.11 divided by 0.11, or about 10.

A decrease in EROI would mean that gross energy extraction must increase; for example, to obtain 1 unit of practical energy at an EROI of 5 requires investing 0.25 unit, for a gross consumption 1.25 units. The situation becomes increasingly desperate at very low EROI values (see Figure 15.6). At EROI of 2.0, procuring 1 unit of useful energy requires the consumption of 2 units of gross energy. At an EROI of 1.5 the gross energy requirement increases to 3 units! At values below this the gross energy requirement quickly rises toward infinity, until finally there is no more useful energy to be had.

The debates over the energy return ratios of corn ethanol or oil shale therefore are not merely academic; small differences between relatively low EROI values can make a huge difference in the total amount of energy consumed. Furthermore, these differences also affect the supporting resources consumed and the potential for associated environmental damage. For example, producing ethanol from corn at an EROI of 1.5 requires, in principle, twice as much land, fertilizer, and water as does ethanol produced from sugarcane at an EROI of 3.0. The same multiplier effect also applies to total emission of CO_2. Taking a more positive view, scientific research aimed at improving the EROI of marginal energy sources could pay large environmental dividends.

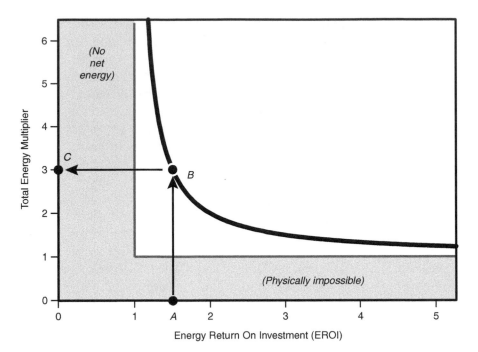

Figure 15.6. Influence of energy return on investment (EROI) on total energy required. Trace upward from an EROI value (point A) to intersect the curve (point B), then to the left to the vertical axis (point C) to find the total energy required to produce 1 unit of useful energy. In the worked example, an EROI value of 1.5 gives a total energy multiplier of 3, meaning that 3 units of total energy must be expended to yield 1 unit of useful energy. Note that no net energy is available at EROI of 1.0 or less, because more energy is consumed than is produced. Total energy multipliers of less than 1.0 are impossible, because this would violate the first law of thermodynamics (i.e., energy cannot be created out of nothing).

Fueling the 21st Century

The days of easily accessible Mesothermic energy appear to be drawing to a close, but the Paleothermic era (founded on coal) lives on, along with the potential for generating larger amounts of nuclear energy. The Cenothermic era of more expensive but apparently abundant fossil fuels is just starting. Are the currently proven reserves of nonrenewable energy (fossil fuel plus nuclear) adequate to take us through the end of the 21st century? Depending on your point of view this landmark could either be viewed as an important goal or as an environmental catastrophe waiting to happen.

Figure 15.7 summarizes estimated reserves and resources of various nonrenewable energy supplies, along with their estimated EROI values. Also shown for comparison are estimates of cumulative world energy consumption in 2000, 2050, and 2100. The latter two curves might be thought of as goals that we hope to meet, based on the assumptions that population will keep growing and that everyone will want a standard of living that is similar to that enjoyed in the developed world today. Let's assume for the sake of argument that over the long term, the different energy sources shown in Figure 15.7 are functionally equivalent. For example, cars that today require liquid fuel could, over time, be converted to use natural gas or electricity. In fact some already do. If this assumption is made, then the contributions from different nonrenewable supplies can simply be added up and compared to projected future energy consumption.

All of the estimates depicted in Figure 15.7 embody significant uncertainties, but some general conclusions can still be drawn. First, if you simply add up all the proven reserves of fossil fuels and uranium, the total falls well short of what is needed to reach 2100. Worse still, the low EROI values associated with oil shale and heavy and extra-heavy oil effectively increase the total amount of energy required by 2100, due to the input or "waste" energy that must be expended.

As noted earlier however, proven reserves do not represent a finite limit. The combination of reserves+resources depicted in Figure 15.7 easily gets us to 2100, and perhaps even through the following century. Whether this rosy scenario will actually play out remains to be seen. Two aspects of the resource estimates deserve special mention however. First, coal presently serves as a low cost fuel; the proven reserves represent only a tiny part of the coal that is known to exist underground. Increased energy prices therefore could result in disproportionately large increases in coal reserves, compared to other fossil fuels. Second, ^{235}U presently accounts for most current nuclear power generation, but represents only 0.7% of natural uranium ore. The technology already exists to use the other 99.3%, which consists of ^{238}U. Doing so would make available vastly greater energy supplies, with very little need of new geological exploration.

Looking even further ahead, several other potentially large energy sources could come into play, such as uranium from seawater, thorium-232 (^{232}Th), nuclear fusion fuels, methane hydrate, and natural gas dissolved in deep brines. They have been omitted from the summary in Figure 15.7 because they have not yet reached the point of commercial exploitation. Their future economic potential may be

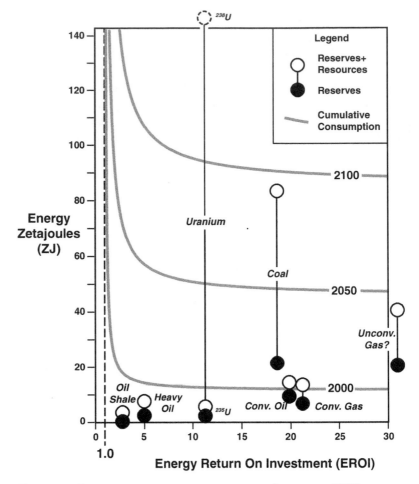

Figure 15.7. Energy content versus energy return on investment (EROI)
for selected nonrenewable energy sources.One zetajoule equals 10^{21} joules
(= one billion trillion joules, the approximate energy content of 175 billion
barrels of oil). The gray curves represent estimated cumulative world energy
consumption from all sources, through the 21st century. These curves slope
upward to the left because of the increasingly large total energy expenditures
required to satisfy net energy demand at lower EROI values (see text for further
explanation). Reserves are known quantities that can be profitably extracted
at present prices and using present technology. Resources are additional
quantities that may become reserves in the future but are subject to large
uncertainties. EROI estimates are also subject to large uncertainties. Values
lying outside the graph area are not plotted to scale (principal data sources
include the BP Statistical Review of World Energy, the U.S. Geological Survey,
Cleveland et al., 1984, Gupta and Hall, 2011, Rogner et al., 2013, Sell et al., 2011,
and the author's estimates).

speculative, but their very existence makes it is clear that the base of the nonrenewable energy pyramid is very broad indeed.

One Year at a Time: The Renewable Alternatives

If we are not going to run out of fossil fuels soon, our best hope for avoiding their negative consequences is to develop other energy sources that can outcompete them. This will not be easy. Coal, oil, and natural gas represent a kind of energy savings account, that allows us to cash in today on energy that accrued over long stretches of the geologic past. In contrast, renewable energy sources require that we live within our present means. Renewable energy is by definition inexhaustible, but the amount that may be obtained within a single year is finite.

Figure 15.8 presents a summary similar to that in Figure 15.7, except that the vertical axis measures power rather than energy. It therefore expresses the capacity of renewable power systems to yield energy at the *rates* that we require. Presently developed renewable energy systems are simply not up to the job. The sum of the black circles, which represent present delivery of renewable power, represents only a small fraction of the world's total power usage today. The difference is being supplied from nonrenewable sources. While it is true that the usage of renewable energy has grown very rapidly in recent years, this success in part reflects our rather low expectations of it. The rapid growth of renewable power is possible precisely because it is currently so small.

Can renewable power grow large enough to replace fossil fuels and other nonrenewable sources? The white circles in Figure 15.8 represent estimates of the potential future capacity of each system. Like nonrenewable energy resources, these estimates are subject to large uncertainties, but some general conclusions can again be drawn. Most notably, the technical limits of some renewable power systems preclude their ever making a majority contribution on their own. For example, river networks concentrate precipitation falling across a broad area into a narrow stream, giving hydroelectric power a high EROI. Favorable topographic conditions are required to take advantage of this natural leverage, however, and most of the best sites are already taken. Likewise, the practical use of waves and tides depends on natural concentration by favorable coastline geometries, and such coastlines are limited in extent. These naturally concentrated renewables therefore represent the tip of the "power pyramid"; their quality is high but their ultimate capacity is relatively small.

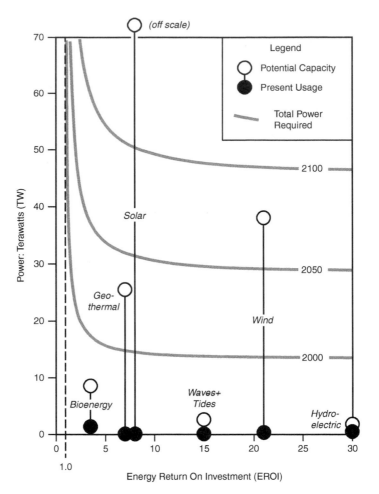

Figure 15.8. Power available versus Energy Return on Investment (EROI) for selected renewable sources.One terawatt equals 10^9 watts (= one trillion watts, equal to approximately three times the rate of energy consumption of the United Kingdom, or one third that of the United States). The gray curves represent estimated world power demand from all sources, through the 21st century. These curves slope upward to the left because of the increasingly large total power required to satisfy net power demand at lower EROI values (see text for further explanation). Potential capacity represents additional power that may be technically feasible to implement in the future, disregarding economic feasibility. Potential capacity and EROI values are both subject to large uncertainties. Values lying outside the graph area are not plotted to scale. (Principal sources include the BP Statistical Review of World Energy, Cleveland et al., 1984, Gupta and Hall, 2011, and Rogner et al., 2013).

The raw amounts of wind, geothermal, and biopower available on the planet are much larger, and these power sources may have considerably more room to grow. The cost of such growth has not been considered in these estimates, however. Early developments of wind and geothermal power have focused on the most favorable geographic occurrences of each. Future growth will eventually have to exploit less favorable sites, just as the petroleum industry has had to venture into deeper waters and more remote locations. The practical potential of biopower remains especially uncertain, in part because its technology is evolving rapidly. Corn ethanol may have very low EROI, but future technologies employing cellulosic feedstock are expected do better. Biopower must also compete with food production for limited soil, water, and fertilizer resources, however. In practical terms, these resources are often no more renewable than fossil fuels.

Solar energy constitutes the base of the renewable power pyramid. It is vastly abundant and appears to be the only renewable power source capable of meeting future demands entirely on its own. It is also a diffuse power source, though, and therefore relatively costly to exploit. According to a 2012 report by Lawrence Berkeley National Laboratory, average solar power installation costs in the United States at that time ran in the range of $5–$6 per watt. The report notes that these costs have been consistently falling, but the total cost of powering the world solely using sunlight would nonetheless be staggering. Even at $2 per watt, installing the 28 terawatt capacity needed by 2050 would cost $56 trillion. This number rivals the current gross world product, which is the total value of all the goods and services produced in a single year.

Conclusions

Many energy issues amount to a mismatch between the timing of energy use versus natural timescales of energy availability. For example, human power demand and the availability of solar power both follow daily cycles, but unfortunately these cycles do not always coincide. Expanded use of solar energy therefore implies the need for some form of short-term energy storage, in order to synchronize supply and demand. At the other extreme, fossil fuels contain energy that has been stored for millions to hundreds of millions of years. They represent a sort of natural battery that can be tapped at will, making daily or seasonal fluctuations in power demand all but irrelevant. We are tapping this battery at rates vastly greater than its natural recharge

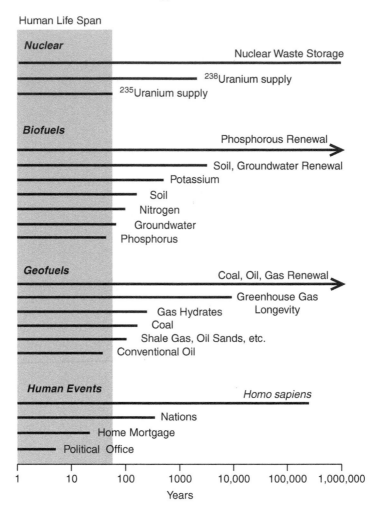

Figure 15.9. Schematic illustration of the duration of various events related to energy resources and consumption.Note that each tick on the horizontal axis increases the time span by a factor of 10. The actual length of each of the bars shown is uncertain and is therefore meant only to illustrate approximate relative timing.

rate, though, raising the prospect of its eventual demise. We will never truly exhaust the Earth's endowment of fossil fuels, but there may come a time when they become economically unattractive.

Precisely when this time will arrive is difficult or impossible to know, because it depends on future economic and technological events that have not yet happened. Past predictions of imminent disaster have failed with remarkable consistency. Current predictions of the future availability of nonrenewable energy vary widely, depending on the energy source, how it is used, and who does the predicting (Figure 15.9). What is not in doubt is that the consequences of energy consumption will be with us for much longer periods, which could well exceed the lifetime of our species. The release of greenhouse gases may peak within the next century or two, but their long-term impact is expected to persist for tens of thousands of years beyond that. Nuclear wastes will follow a similar pattern; the greatest impact will occur within the first century as fission products decay, but lower levels of radioactivity will continue for many thousands of years thereafter.

Biofuels production combines temporal aspects of both renewable and nonrenewable energy systems. Crop growth is subject to seasonal cycles, but fuel storage and global fuel distribution systems could in principle be designed to ameliorate the negative impacts of these cycles. Biofuel production also requires other resources that have timelines that resemble those of fossil fuels. Soil and groundwater depletion have already become significant problems in some areas, and it has been suggested that high-quality phosphorus deposits could be depleted within a century. Once gone, replenishment of these resources will require hundreds to millions of years.

For More Information

Adelman, M. A., and Lynch, M. C., 1997, Fixed view of resource limits creates undue pessimism: *Oil & Gas Journal*, v. 95, p. 56–60.

Anonymous, 1950–1964, Statistical report on world's reserves, production, refining: *Oil & Gas Journal*, volumes 48–63.

Archer, D., 2005, Fate of fossil fuel in geologic time: *Journal of Geophysical Research*, v. 110, doi:10.1029/2004JC002625.

Barbose, G., Darghouth, N., and Wiser, R., 2012, Tracking the Sun. V: An Historical Summary of the Installed Price of Photovoltaics in the United States from 1998 to 2011: Lawrence Berkeley National Laboratory Report 5919E, 61 p.

BP, 2013, BP Statistical Review of World Energy: available online at bp.com/statisticalreview.

Campbell, C. J., and Laherrere, J. H., 1998, The end of cheap oil: *Scientific American*, v. 278, p. 78–83.

Cleveland, C. J., 1984, Costanza, R., Hall, C. A. S., and Kaufmann, R., 1984, Energy and the U.S. economy: A biophysical perspective: *Science*, v. 225, p. 890–897.

Deffeyes, K. S., 2001, *Hubbert's Peak: The Impending World Oil Shortage*: Princeton, NJ, Princeton University Press, 208 p. 1–26.

Duncan, R. C., 2001, World energy production, population growth, and the road to Olduvai Gorge: *Population and Environment*, v. 22, p. 503–522.

Dyni, J. R., 2006, Geology and Resources of Some World Oil-Shale Deposits: *United States Geological Survey Scientific Investigations Report* 2005-5294, 42 p.

Gorelick, S. M., 2009, *Oil Panic and the Global Crisis: Predictions and Myths*: London, Wiley-Blackwell, 256 p.

Gupta, A. K., and Hall, C. A. S., 2011, A review of the past and current state of EROI data: *Sustainability*, v. 3, p. 1796–1809.

Intergovernmental Panel on Climate Change, 2014, Climate Change 2014: Impacts, Adaptation, and Vulnerability: Cambridge, Cambridge University Press, 76 p.

Johnson, R. C., Mercier, T. J., and Brownfield, M. E., 2011, Assessment of In-Place Oil Shale Resources of the Green River Formation, Greater Green River Basin in Wyoming, Colorado, and Utah: *United States Geological Survey Fact Sheet* 2011-3063, 4 p.

Lewis, N. S., 2007, Powering the planet: *Materials Research Society Bulletin*, v. 32, p. 808–820.

Masters, C. D., Attanasi, E. D., and Root, D. H., 1994, World petroleum assessment and analysis, *in* Proceedings of the 14th World Petroleum Congress, Chichester, UK, John Wiley & Sons, v. 5, p. 529–541.

McCabe, P. J., 1998, Energy resources – cornucopia or empty barrel?: *American Association of Petroleum Geologists Bulletin*, v. 82, p. 2110–2134.

McCabe, P. J., 2012, Oil and natural gas: Global resources, *in* Meyers, R. A., ed., Encyclopedia of Sustainability Science and Technology: Springer Science & Business Media, 13 pp.

McCormick, M., Büntgen, U., Cane, M. A., Cook, E. R., Harper, K., Huybers, P., Litt, T., Manning, S. W., Mayewski, P. A., More, A. F. M., Nicolussi, K., and Tegel, W., 2012, Climate Change during and after the Roman Empire: Reconstructing the Past from Scientific and Historical Evidence: *Journal of Interdisciplinary History*, v. 43, p. 169–220.

McKelvey, V. E., 1972, Mineral resource estimates and public policy: Better methods for estimating the magnitude of potential mineral resources are needed to provide the knowledge that should guide the design of many key public policies: *American Scientist*, v. 60, p. 32–40.

Meyer, R. F., and Attanasi, E. D., 2003, Heavy Oil and Natural Bitumen – Strategic Petroleum Resources: *U.S. Geological Survey Fact Sheet* 70-07, 5 p.

OECD Nuclear Energy Agency and the International Atomic Energy Agency, 2012, *Uranium 2011: Resources, Production and Demand*: Paris, OECD/NEA, 486 p.

Rogner, H. H., Aguilera, R. F., Archer, C. L., Bertani, R., Bhattacharya, S. C., Dusseault, M. B., Gagnon, L., Haberl, H., Hoogwijk M., Johnson, A., Rogner, M. L., Wagner, H., Yakushev, V., Arent, D. J., Bryden, I., Krausmann, F., Odell, P., Schillings, C., and Shafiei, A., 2013, Energy resources and potentials, *in GEA, 2012: Global Energy Assessment – toward a Sustainable Future*: Cambridge and New York, Cambridge University Press, and the International Institute for Applied Systems Analysis, Laxenburg, Austria., p. 425–512.

Sell, B., Murphy, D., and Hall, C. A. S., 2011, Energy return on energy invested for tight gas wells in the Appalachian basin, United States of America: *Sustainability*, v. 3., p. 1986–2008.

Simmons, M. R., 2006, *Twilight in the Desert: The Coming Saudi Oil Shock and the World Economy*: Hoboken, NJ, John Wiley & Sons, 464 p.

Smith, George Otis, 1921, A foreign oil supply for the United States: *Transactions of the American Institute of Mining and Metallurgical Engineers*, p. 89–93.

Tainter, J. A., and Patzek, T. W., *Drilling Down: The Gulf Oil Debacle and Our Energy Dilemma*: New York, *Copernicus*, 251 p.

U.S. Geological Survey World Energy Assessment Team, 2000, U.S. Geological Survey World Petroleum Assessment 2000 – Description and Results: *U.S. Geological Survey Digital Data Series* 60, four CD-ROMs.

U.S. Geological Survey World Conventional Resources Assessment Team, 2012, An Estimate of Undiscovered Conventional Oil and Gas Resources of the World, 2012: *U.S. Geological Survey Fact Sheet* 2012–3042, 5 p.

Yergin, D., 1992, *The Prize: The Epic Quest for Oil, Money, and Power*: New York, Touchstone, 885 p.

Yergin, D., 2011, *The Quest: Energy, Security, and the Remaking of the Modern World*: New York, Penguin Press, 804 p.

16
Conclusions

What's past is prologue.

William Shakespeare, The Tempest

It should be clear by now that the past geological evolution of the Earth has directly influenced the present and future availability of *all* energy resources, in one way or another. This connection is obvious for fossil fuels, nuclear, and geothermal energy, all of which exploit the tremendous natural leverage of geologic time. In contrast, renewable systems rely solely on contemporary energy fluxes, which are very large in their own right. The evolution of the Earth has literally set the stage for their use, however, by shaping the geography of practical renewable energy. It has also predetermined the availability of nonrenewable Earth resources that are required to obtain renewable energy.

This dependency on the geologic history of the Earth places intrinsic limitations on all energy systems. Note that these limitations are not static, but instead evolve continuously through time. Three overarching principles govern the changing relationships among humans, energy, and Earth resources, as detailed in the following.

Quality, Not Quantity

Our voracious energy appetite has stimulated spirited debate over the ultimate quantities available from different sources, motivated by concerns that supplies might eventually fall short of demand. In a gross sense this concern is misplaced, however. For example, the total amount of solar power reaching the Earth exceeds current human power consumption by a factor of more than 10,000. The total amount of energy stored in organic matter buried in the Earth's crust is on the order of five thousand times greater than the presently recognized magnitude of fossil fuels. Various other energy systems hold similarly mind-boggling surpluses. The Earth clearly has plenty of energy. What matters therefore is not quantity, but quality. Energy quality can be defined in many ways; in the present context it represents the relative potential for doing useful work, at a minimal cost.

Basic thermodynamic considerations dictate that highly concentrated energy sources can be exploited more efficiently than more diffuse ones and therefore represent higher quality. The primary reason that fossil fuels have proven so revolutionary is that they naturally concentrate sunlight, a diffuse resource. This process of natural concentration has occurred partly in time, through the accumulation of solar energy over many millions of years. It has also occurred in space, through the localized enrichment of combustible organic deposits. Other nonrenewable resources have experienced comparable processes of natural concentration. For example, minable uranium ores contain at least one part uranium per thousand by weight, a natural enrichment of 5 million times compared to uranium's average terrestrial abundance. Geothermal energy developments have generally focused on areas where the temperature of near-surface rocks has been greatly increased by the localized ascent of hot magma.

Renewable energy resources also depend heavily on natural concentration. Hydroelectric dams provide perhaps the most striking example of this. They work by focusing the gravitational energy of precipitation falling across an enormous catchment area, into a single narrow river course. It is no accident that hydroelectric was the first large-scale renewable energy system to be developed, and it remains important today. Wind turbines also exploit natural concentration; they are most productive where near-surface winds naturally blow at velocities greater than average, such as ridge tops, unobstructed plains, or windy stretches of coastline.

Bioenergy represents something of a unique case in that it does not initially appear to benefit from any natural amplification; for example, the conversion of raw feedstocks into concentrated liquid fuel must be accomplished artificially in a refinery. However, the soils needed to grow those feedstocks contain nutrients that have been naturally concentrated over long periods, within a thin surface layer above their parent rocks. Artificial fertilizers needed to replace these nutrients result mostly from naturally concentrated, geologic deposits.

Naturally concentrated, high-quality energy resources represent only a tiny part of the Earth's total energy endowment, a principle commonly expressed through the metaphor of the resource pyramid. The volume of the pyramid represents the Earth's total endowment of a particular resource, whereas concentrated, high-quality resources are represented by the tip. Exponentially increasing amounts of energy can be had by digging deeper into the pyramid, but that energy is obtained at ever-increasing costs. The important question is not the size of the pyramid, but how far down we can afford to dig. We will run out of money long before we run out of energy.

This same basic logic applies equally well to both nonrenewable and renewable resources. For example, the magnitude of concentrated power that may be theoretically obtained from hydroelectric dams or wind turbines is a minuscule fraction of the total energy contained in raindrops or the wind. The useful potential of renewable energy depends on prevailing technology and on energy prices, but generally speaking the highest-quality, most naturally concentrated resources are among the first to be exploited. At some point the continued growth of renewable energy output will depend on exploiting lower-quality natural resources. Assuming no change in the underlying technology of these systems, the cost per unit of energy obtained might be expected to increase.

Technology Changes Reality

All predictions are by necessity rooted in present reality, for the simple reason that we cannot really know the future. However, if the recent past is any guide we can be reasonably certain that the future will include change, and that new technologies will fundamentally alter our perception of the available energy resources.

Prior to the 1850s crude oil was largely a geological curiosity, used for such miscellaneous purposes as road paving, the formulation of patent medicines, or the caulking of wooden ships. Its potential value as an energy source was largely unknown, and known reserves of crude oil were practically nonexistent compared to those today. Since then the technologies for finding, extracting, refining, and utilizing crude oil and natural gas have advanced continuously. Our perceptions of the economic magnitude of these resources have grown apace. Most recently, the application of horizontal drilling and hydraulic fracturing has allowed us to extract large amounts of oil and gas from rocks that most geologists believed to be impervious to such efforts as recently as the 1990s. These efforts have created large new reserves of Cenothermic fossil fuels, which previously did not exist.

The idea of extracting energy from the nucleus of an atom would have been unthinkable prior to the very last years of the 19th century, and unworkable until the middle of the 20th. It has since become commonplace, and the main limitations on nuclear energy are now imposed by public policy and the availability of investment capital, rather than technology. Renewable energy technologies have also advanced markedly since the mid-18th century, primarily as an outgrowth of the invention and refinement of practical dynamos that can convert the motion of water and wind directly into electricity. The transformation of sunlight directly into electricity without use of dynamos was impossible prior to the invention of the first practical solar photovoltaic cell, developed at Bell Labs in 1954. Solar collectors employing this technology are now cheaply available at any local hardware store, and the cost of manufacturing them continues to drop.

Similar innovations can reasonably be expected to continue in the future. Fossil fuel extraction technologies are evolving rapidly, and several different approaches to mitigating their atmospheric impact are also being explored. Technologies for solar collection and storage are also advancing rapidly, as is the quest for commercial biofuel production from cellulosic feedstocks. Other known opportunities for advancement include biofuels produced using algae, offshore wind development, expanded use of geothermal energy, and nuclear fusion. Technologies for transmitting, storing, and conserving energy are also improving. It is difficult to predict which innovations will bear the most fruit, but it seems inevitable that new energy supplies *will* be made available.

There Is No Free Lunch

All potential energy sources present implicit trade-offs between the benefits they provide and their various environmental, financial, or social costs. This is perhaps most obvious for fossil fuels. Elevated atmospheric CO_2 concentration was first proposed to cause atmospheric warming by geologist Thomas Crowder Chamberlin in 1899, in an effort to explain the coming and going of ice ages. The realization that human activities could alter this balance dawned much later, long after fossil fuels had become our dominant energy source. It is now plainly understood that the unprecedented rise in global prosperity associated with fossil fuel combustion has a steep environmental price. Emissions of CO_2 have already altered the greenhouse gas composition of the atmosphere, and climate scientists expect that resultant environmental degradations will continue for centuries. This knowledge has served to heighten public awareness of the full costs of energy consumption, an important step toward positive change. However, it can also lead to the illogical conclusion that if fossil fuels are bad, then any alternative must by definition be good (or at least better).

What's missing from this view is the realization that all of the alternatives have costs of their own, and that those costs will inevitably scale up in size according to the amount of fossil fuel consumption that is displaced. Hydroelectric and nuclear are currently the leading alternatives to fossil fuels, and not surprisingly they also tend to attract the most attention to their downsides. Biofuels are also drawing increasing amounts of criticism, in part because of the large energy investments required to produce ethanol from corn. Even if produced from cellulosic feedstocks, it seems unlikely that biofuels can ever approach the energy return ratios of Mesothermic fossil fuels. Furthermore, biofuel production will always hold the potential to compete with food production, since both require the same supporting geologic resources. As bad as the effects of global warming may become over the next century, famine and malnutrition are by nature more immediate problems.

Renewables such as wind, solar, and geothermal have made more modest contributions to date and generally appear more environmentally benign. Perhaps they are, but until they grow significantly in scale it is hard to know for sure. One hundred years ago no one dreamed that the combustion of fossil fuels might result in the melting of polar ice. Substantially increased use of solar, wind, wave, or geothermal energy also appears likely to require massive new investments in infrastructure, both for primary collection and for secondary transmission and storage. The need to make such investments in order to sustain our growing energy use will likely make a direct impact on the world economy.

In the end the only truly "green" energy strategy is to use less of it. This will not be so easy to do, however, because energy use is closely associated with wealth. Most of us in the developed world probably do not want to return to the horse-and-buggy lifestyles experienced by our ancestors only a few generations ago. Since we cannot escape the negative consequences of our own energy consumption, we will be forced instead to choose the consequences we deem least damaging.

Index

Abiogenic origins
  natural gas, possibility of, 136–137
  oil, possibility of, 135–136
Absorption, effect on solar energy,
  36–37
Adelman, M.A., 329
Adsorption of methane, 155–156
Aesthenosphere, 17
Agriculture
  algae resulting from runoff,
    116–117
  amount of land dedicated to,
    106f6.2
  erosion of soil and, 112–114
  eutrophication resulting from
    runoff, 116–117
  fertilizers (See Fertilizers)
  forests, conversion to farmland,
    114–115
  grasslands, conversion to
    farmland, 114–115
  groundwater, use of, 251
  historical background, 105–107
  hypoxia resulting from runoff,
    116–120
  irrigation in, 251, 254f12.7
  plowing, effect of, 112, 114
  soil and, 90–91
  water, use of, 251
Alcohol. See Ethanol
Algae
  from agricultural runoff, 116–117
  biofuels from, 100–101
  fossil fuels from, 124
Alifisols, 92
Alpha particles, 264–265, 266
Animals, fossil fuels from, 123, 124

Anthracite coal, 127
Anticline, 171, 171f9.4, 180
Appalachian Basin, 175
Appalachian Mountains, 145
Aquifers
  depletion of, 8–9
  freshwater aquifers, avoiding
    carbon sequestration in,
    295–297
  High Plains Aquifer, 251–255
  hydraulic fracturing and
    freshwater aquifers, 208f10.7
  saline aquifers, carbon
    sequestration in, 294–298
Archaea
  gas hydrates, role in formation of,
    215–216
  oil sand, role in formation of,
    215–216
Asphaltenes, 217–219
Aswan High Dam, 72
Atacama Desert, 45–46
Atchafalaya River, 117, 120
Atchafalaya Swamp, 124
Athabasca Basin, uranium in,
  281–282
Athabasca River region, oil sand in,
  220–224
Atmosphere. See also Wind energy
  carbon dioxide in, 26–28
  condensation, latent heat from,
    69–71
  convective air masses, 60–61
  Ferrel cells, 59
  formation of, 14
  geology, effect on, 26–28
  Hadley cells, 59